T0092080

HIDDEN IN PLAIN SIGHT

HIDDEN IN PLAIN SIGHT

The History, Science, and Engineering of Microfluidic Technology

ALBERT FOLCH

The MIT Press
Cambridge, Massachusetts
London, England

© 2022 Massachusetts Institute of Technology

All rights reserved. No part of this book may be reproduced in any form by any electronic or mechanical means (including photocopying, recording, or information storage and retrieval) without permission in writing from the publisher.

The MIT Press would like to thank the anonymous peer reviewers who provided comments on drafts of this book. The generous work of academic experts is essential for establishing the authority and quality of our publications. We acknowledge with gratitude the contributions of these otherwise uncredited readers.

This book was set in Scala Pro and Scala Sans Pro by Westchester Publishing Services. Printed and bound in the United States of America.

Library of Congress Cataloging-in-Publication Data

Names: Folch i Folch, Albert, 1966– author.
Title: Hidden in plain sight : the history, science, and engineering of microfluidic technology / Albert Folch.
Description: Cambridge, Massachusetts : The MIT Press, 2022. | Includes bibliographical references and index.
Identifiers: LCCN 2021033924 | ISBN 9780262046893 (hardcover)
Subjects: LCSH: Microfluidics. | Microfluidic devices. | Biomedical devices.
Classification: LCC TJ853.4.M53 F65 2022 | DDC 620.1/06—dc23
LC record available at https://lccn.loc.gov/2021033924

10 9 8 7 6 5 4 3 2 1

To Mehmet Toner—as a triple token for his mentorship, generosity, and friendship

CONTENTS

PROLOGUE

I am a professional scientist, and I wanted to explain to people what my field—microfluidics—is, how it works, and how it came to be. "What is microfluidics?" you are likely to ask. In contrast, everyone knows what microelectronics is: the marvelous technology inside smartphones, televisions, and videogames. It's easy to explain microelectronics: it's electronics, but so small you can't even see it.

Microfluidics is a miniaturization technology related to microelectronics, but applied to fluids instead of electricity. Microfluidic technology is like all the pipes and valves that allow for routing fluids through your house and your car, but so tiny you can barely see the channels with the naked eye. Microfluidics may be less celebrated than microelectronics, but we build them with similar technology, and you use microfluidics just as much. Microfluidics underlies vital things—an inkjet printer, a glucometer, a pregnancy test, a DNA sequencing chip, and so many other devices that researchers are developing to improve your life. You might not have known that these devices existed because engineers purposely hid them under a user interface, the same way you cannot touch the microelectronics of your smartphone or your car. They are hidden in plain sight.

I have always believed that you cannot fully grasp a concept without understanding the process of its inception. Why are microfluidic devices typically made in a single material? Why are they expensive to make? We learn about an invention through a final publication, but before that date, there has been a long road of brainstorming, adding team members, raising funds, writing grants, and submitting patents and manuscripts. In 1905, a patent clerk named Albert Einstein published his *theory of special relativity*, in which he demonstrated the contraction of time imagining a clock on a train compared with that on a train station.[1] Einstein's work is one of the most famous achievements in the history of science. But perhaps less known is the fact that, around that period, Einstein was examining the merits of a patent on light-synchronized clocks for improving railroad scheduling.[2] This point serves to illustrate that discoveries and inventions are not conceived in a vacuum. In my book, I purposely picked the applications, the stories, and even the chapter order to give the reader a glimpse of the history of microfluidics through each invention because—in the wise words of the sixth US president, John Quincy Adams—"who we are is who we were."[3] Hence, this book is my attempt to unveil not only the inner workings of the devices that are hidden in plain sight but also the efforts, teams, places, and circumstances that enabled those inventions and that, once archived in memory, end up below the surface as well. Only by learning the history of microfluidics can one fully understand its materials and manufacturing challenges.

I intend this book for people with some background and interest in science but no background in microfluidics—someone who might not even have heard about the term *microfluidics*. If the book raises the awareness of microfluidics among a wider audience, I will have done my job. In the following pages, you will learn the stories of how inventors developed microfluidic devices and turned them into products—or struggled to do so. You will "hear" the stories from the engineers themselves. You will "meet" the engineers behind these inventions and walk with them on the discovery path to learn what challenges they faced, what solutions they adopted, and what eureka moments they enjoyed.

This book does not pretend to be an overview of the microfluidics field or a who-is-who in microfluidics. Thus, necessarily, a lot of excellent and critical efforts are missing. To write the book, I picked a few star applications and used them to build narratives with one or two researchers for each. Although in general earlier work is presented first, the narrative of the different applications necessarily overlaps in time—the chapters are not linearly concatenated in time. Most importantly, I'd like to apologize to all my microfluidic colleagues who might have expected that their efforts surely deserved mention in a book about microfluidics. The narrative had to be slim, so there was no space to mention the research of microfluidic giants such as Chong Ahn, Nancy Allbritton, Albert van den Berg, Jeffrey Borenstein, Mark Burns, Yoon-Kyoung Cho, Nikos Chronis, Utkan Demirci, Tejal Desai, Pat Doyle, Katherine Elvira, Teruo Fujii, Kevin Healy, Sarah Heilshorn, Amy Herr, Tony Jun Huang, David Issadore, Klavs Jensen, Shana Kelley, Ali Khademhosseini, Michelle Khine, Catherine Klapperich, Wilbur Lam, Blanca Lapizco-Encinas, Thomas Laurell, Gwo-Bin Lee, Luke Lee, Andre Levchenko, Dorian Liepmann, Susan Lunte, Scott Manalis, Ellis Meng, Hywel Morgan, Aydogan Ozcan, Sumita Pennathur, Jonathan Posner, Beth Pruitt, Jianhua Qin, Michael Ramsey, Juan Santiago, Amy Shen, Lydia Sohn, Roman Stocker, Howard Stone, James Sturm, Patrick Tabeling, Shoji Takeuchi, Joe Tien, Hsian-Rong Tseng, Victor Ugaz, Joel Voldman, Steve Wereley, Mingming Wu, and Huabing Yin, to list a few friends and dear colleagues—and many, many others. Yet their creativity and impact are beyond doubt. The detection of cells and molecules with microfluidic devices using acoustic forces (acoustofluidics), light (optofluidics), and magnetic fields, or the investigation of worms and bacteria, to name a few pivotal areas, have been left out. For this reason, the book should not be read as a thorough review of the field. For each development or discovery, the date typically refers to the publication year; this date should only be taken approximately, as devices and findings are the result of long quests peppered with several peer-reviewed manuscripts, presentations at conferences, and patent applications. Including all these satellite references would have more than doubled the length of the bibliography and wasted too many trees.

To build the stories that you will read in the book, I had to contact many people. Luckily, at the time of writing this book, these stories, these developments, these actors were just one email or Zoom call away—I have known most of them for a long time, and I feel fortunate that there was only one or at most two degrees of separation between them and me. I am indebted to the following individuals from the microfluidics community (in alphabetical order) for sharing valuable information in the form of personal and professional memories, suggestions, and/or images during multiple emails, Zoom interviews, and/or LinkedIn messaging exchanges: Rashid Bashir, Holger Becker, Hans Biebuyck, Friedrich Bonhoeffer, Michael Breadmore, Manoj Chaudhury, Daniel Chiu, Brian Cunningham, Emmanuel Delamarche, Andrew deMello, Dino Di Carlo, David Duffy, Felice Frankel, Will Grover, Adam Heller, Steve Higgins, Dan Huh, Don Ingber, Hal Jerman, David Juncker, Greg Kellogg, Amit Kumar, Jennifer Lewis, Vincent Linder, Laurie Locascio, Matthias Lutolf, Brendan D. MacDonald, Marc Madou, Andreas Manz, Andrés Martínez, Richard Mathies, Alec Mian, Arman Naderi, Greg Nordin, Nicole Pamme, Philippe Renaud, John Rogers, Holger Schmidt, Olivier Schueller, Sam Sia, Alison Skelley, Steve Terry, Mehmet Toner, Elfi Töpfer, Marc Unger, David Weitz, Aaron Wheeler, and Adam Woolley. Through them, I learned incredible stories that I'm honored to share here with you. Understandably, many others did not have time to answer my request. (Only a few women answered, and I wish more women out of the regrettably male-dominated pool had answered; I cite the work of these women pioneers in the book.) The inquisitive reader will realize that many of these stories intersect through a remarkably small number—about three—of institutions and subfields—also two or three. An analogy I like to make is that our tightly knit community of engineers is a bit like that of the Formula 1 engineers: we are competing against each other with our designs but as soon as one of us finds a clever feature that improves performance, it is rapidly celebrated and adopted. It took a village to raise this kid.

The images that you see on the pages tell as much as the words. The field of microfluidics produces stunning imagery, and—perhaps for that reason—more than a few microfluidic engineers have become talented photographers. I am indebted to Sara Stearns and the Chemical and Biological Systems Society (CBMS) for letting me use an extensive collection of beautiful photographs of microfluidic devices that were taken by Niall McDonald (Okaar Photography) for an exhibit at the MicroTAS 2016 conference in Dublin. I also emailed a large number of colleagues requesting "beautiful images"; so many answered that I could not possibly include them all, but my deep thanks go to all who answered just for the privilege of being shown their best images. I must also collectively thank my whole group for producing a wealth of gorgeous photos that inspired me to start BAIT in 2007. BAIT—short for Bringing Art Into Technology—is an outreach art program whose aim is to entice nonscientists to become interested in science using beautiful imagery and movies. I remain convinced that the BAIT images will outlast the science that originated them. Many of the photos you see in the book have been taken by my students and have been framed and exhibited in public halls. Beautiful art can be an effective bait to catch people's attention to science.

I also found some scattered resources—interviews, talks, websites, and journal reviews—that provided additional information on some historical events, and I cite those in the book. The three special *Lab on a Chip* issues on Canada, Japan, and Switzerland were a good starting point because pioneering groups in these three countries played a crucial role in the early history of microfluidics. I wholeheartedly recommend visiting the YouTube Channel The Lutetium Project, where one can find superb introductory videos to the science and technology of microfluidics. The book is biased toward peer-reviewed and otherwise easily accessible sources of information. Thus, I might have left out contributors who published mostly in the form of patents or conference proceedings, neither of which contain enough details and are often not reviewed with the same rigor as journal publications. In the few cases where some proceedings or patents appear in references consistently over the years, I have also used them as a source. I'm very thankful to my friend and Hollywood screenwriter Jeff Spencer for his vivid storytelling of his lifelong experience as a diabetic patient and to my dearest friend and professor Xevi Verdaguer (University of Barcelona) for chemistry consultations.

This book would not have been the same without my wife Lisa Horowitz, who has also become my partner in research. Lisa was trained as a molecular neurobiologist with Nobel laureate Linda Buck and was even invited to the Nobel ceremony, leaving me the terrifying task of taking care of our one-year-old by myself. Perhaps to repay me, Lisa decided to help me develop some of the key concepts driving my lab, in particular our designs focused on user-friendliness and our idea of intact-tissue microfluidics. She is brilliant at finding the logical flaws of any argument—including in our domestic disputes, unfortunately—so logically—no pun intended—I have been quick to use her exceptional skills for improving this book. She read it, edited it from beginning to end, and came up with the clever title. I'm also extremely grateful to my stepfather-in-law David Guenther for such meticulous editing that only a retired programmer could have endured. Yet all mistakes are my fault and nobody else's.

Finally, this book could not have seen the light of day without the altruistic dedication of countless anonymous health-care and delivery workers that made it possible for me to write and do research from the safety and comfort of my privileged home during the COVID-19 pandemic. To all of them, my heartfelt gratitude.

Seattle, April 5, 2021

INTRODUCTION: THE IMAGE THAT CHANGED MY LIFE

As often happens in life, I went into microfluidics due to circumstances that were out of my control. The first was a bit of a misunderstanding by the Massachusetts Institute of Technology (MIT) provost and professor of chemistry Mark Wrighton when he visited the University of Barcelona (UB), Spain, where I was a physics graduate student. From 1990 until 1992, my PhD advisor Javier Tejada, who also played next to me in the midfield of the UB Physics Department soccer team, sent me to the lab of Miquel Salmeron at the Berkeley Lawrence Laboratory to learn *atomic force microscopy* (*AFM*). At Berkeley, I was struck by a lightning called Lisa Horowitz, a dazzling molecular biologist. But she was more allured by Harvard Medical School's MD/PhD than by me, so I became determined to follow her to Boston one day. Back in Barcelona, for my PhD, I made submicron-sized carbon deposits using *scanning electron microscopes* (*SEMs*) and imaged them with an AFM. SEMs consist of an electron beam in a very clean, high-vacuum chamber, but the few molecules of residual contaminants present in the chamber are sufficient to deposit carbon soot where the electron beam strikes the surface. By repositioning the beam, I was able to write carbonaceous "UB" for fun—not knowing that it would change my life later.

But I was missing my brown-eyed girl, so for summer 1992, I used all my convincing powers as a soccer teammate to sway my PhD advisor: we should use the SEM facility at MIT instead—just 1 mile from Lisa's apartment—to do the experiments; it worked. Before the summer, my advisor announced that the provost of MIT was visiting the UB campus for a high-level visit and invited me to the meeting—with the deans and several prominent professors. At the end of the visit, I introduced myself to Mark Wrighton and showed him a picture of my AFM image of tiny

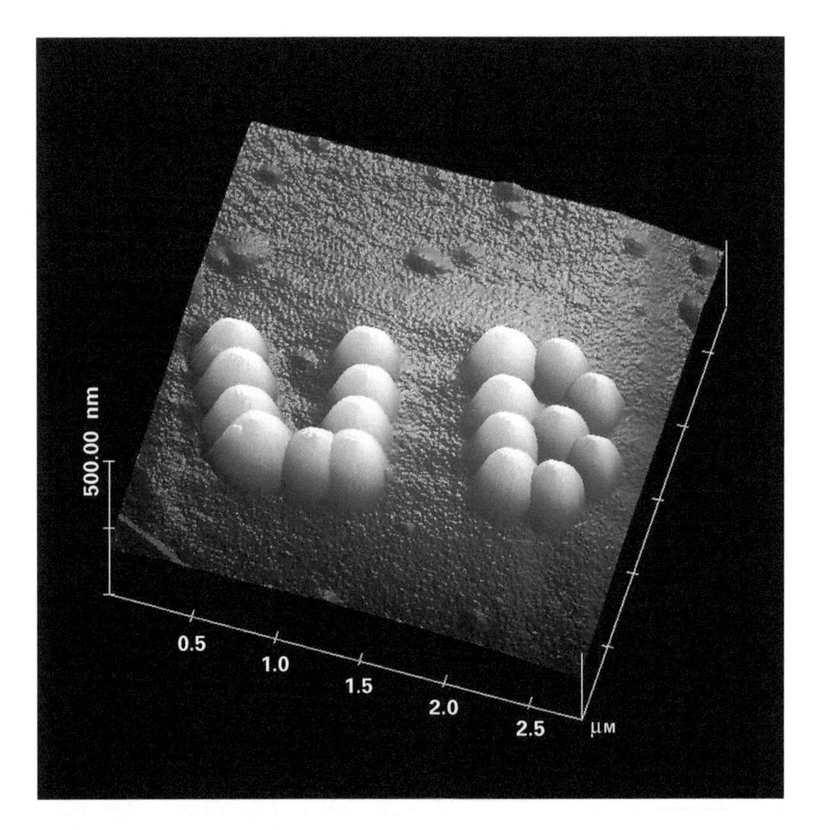

Figure I.1
AFM image of electron-beam-deposited carbon needles.
Image by Albert Folch, then at University of Barcelona.

carbon deposits, which had been false-colored in gold (figure I.1). In his electrochemistry lab, Mark was pursuing the idea of measuring the conductance of a single-polymer molecule, and he was looking for ways to make tiny gold electrodes. I had to clarify that the dots were made of carbon and that the AFM image had been pseudo-colored, but he insisted: "Can you make those deposits with gold?" I quickly remembered a paper that had deposited gold dots using an organometallic compound, so I jumped into the void: "Yes, it should be possible by using dimethyl acetylacetonate gold." Mark Wrighton pouted his lips in his characteristic manner and said: "Good. Come see me when you arrive at MIT this summer."

That summer, Wrighton generously granted me an appointment on my arrival. (To change my visa status, Lisa and I drove to the Montreal Jazz Festival for the weekend and requested a different visa at the US border; I doubt it is possible to do this anymore.) I worked hard from 10 a.m. until 10 p.m. MIT had an environmental SEM, which allowed me to add traces of oxygen or water vapor to the organometallic gas and resulted in higher-purity deposits. When I met with Wrighton to show him the results, he congratulated me and gave me an enduring piece of advice: "Good. Always keep people excited." Wrighton was genuinely impressed that I was able to finish a manuscript (which was later accepted in *Applied Physics Letters*[1]) in just under twelve weeks, so he agreed to take me for a postdoc after I finished my PhD in Barcelona.

I landed in Boston on April 1, 1994, to pursue a project that entailed the microfabrication of AFM cantilevers.[2] I had no microfabrication training, but fortunately Marty Schmidt in the Electrical Engineering and Computer Science Department, a world expert in microelectromechanical systems (MEMS), was my co-advisor. When Mark Wrighton left MIT, I transferred to Marty Schmidt's group. (During the cleanup of the Wrighton lab, we found a vial containing an unidentified chemical labeled "Linus' reagent" with Wrighton's handwriting: Wrighton clarified that it was vitamin C—a humorous reference about how his PhD advisor, the Nobel laureate Linus Pauling, had defended the health benefits of megadoses of vitamin C.) Marty Schmidt and Octavio Hurtado (the MIT clean room manager) turned out to be outstanding teachers; in his office, with just pen and paper, Marty patiently taught me how to design a fab process, and Octavio taught me how to be meticulous, persistent, and safe in the clean room. In 1997, as the end of my fellowship was coming dangerously close, Marty introduced me to a young Harvard professor named Mehmet Toner who was looking for a postdoc and I joined his lab. Soon Toner hired Octavio as well to set up a small clean room for photolithography, which I had to myself. Octavio would still peek in from time to time to correct my technique in his gently, fatherly manner. That bond with Octavio resurfaces every time I correct my students, trying to be like him. Mehmet and I published our first paper in microfluidics in 1998, followed by many others. Like my parents, Mehmet Toner has a deeply intellectual mind. He taught me to pay attention to the biomedical fundamentals first—something that was very dear to my physics background—and to listen before I speak—something I was not good at back then. "The system has a memory," he likes to say. His mentorship was fun and invigorating, and I keep a list of his proverbial advice that resonates in my head from time to time.

I left Toner's lab in 2000 to become a professor at the University of Washington's Bioengineering Department in Seattle. It is from Seattle that I wrote this book dedicated to Mehmet Toner. After my small, initial contribution, Toner developed one of the most prestigious microfluidic centers in the world and became one of the most-cited microfluidic engineers alive. This book is dedicated to him in part for his achievements, but mostly because a smile rushes to my face every time he steps into my mind.

1 TINY CHANNELS EVERYWHERE

MICROFLUIDICS IN NATURE

Microfluidic systems are any device or natural structure that processes minuscule amounts of liquids, usually on a scale that is smaller than a millimeter. These processes—inspiring to researchers—have existed in nature for eons. Humans have been fascinated for generations by the formation and rolling of dewdrops on leaves (figure 1.1), the wicking of dyes into cloth fibers, the seeping of water into beach sand, and the prevention of wetting by a coat of beeswax. Children stop playing to stare in awe at how water striders bend water like a rubber film with their tiny legs (figure 1.2), gracefully sliding on a lake as an ice skater would slide on a frozen pond. We have all looked up a forest tree and wondered how the tallest branches are effortlessly able to pump up, sometimes for centuries and without a beating heart, the moisture from the ground. The ground, made of dirt and mud, is also abundantly microfluidic: as rain infiltrates the soil, nutrients dissolved in water percolate and diffuse through small cracks and are sensed by worms and other creatures as they crawl in the darkness of this porous, highly connected underground realm. The world is rich with microfluidic phenomena.

Figure 1.1
Dewdrops on a leaf.
Image source: Pxhere.com.

Figure 1.2
The surface tension of water allows a water strider to stay afloat.
Image source: Pxhere.com.

We study these phenomena because they help us understand the world around us and, with this knowledge, we can engineer solutions that benefit us. Understanding how water interacts with the objects that surround us allows us to make better detergents and dyes for our clothes, cements that are porous but do not allow for water penetration, and bacteria-repellent coatings for hospital linens and surgical instruments. A mosquito bite may be irritating, but it is also a microfluidic marvel: after locating an ideal spot on your skin, a mosquito drills into that spot with her fascicle (figure 1.3)—a microscopic needle—and injects a minuscule amount of saliva before pumping out her blood meal. The mosquito's fascicle has inspired the design of new and more effective microneedles for delivering drugs through your skin.[1] We put ointments—a special mixture of microfluidic oil droplets in water—on our skin to block or confuse the sense of smell of the mosquito and thus prevent the mosquito from biting us. Scientists study the silk-spinning organs of spiders (called spinnerets) to try

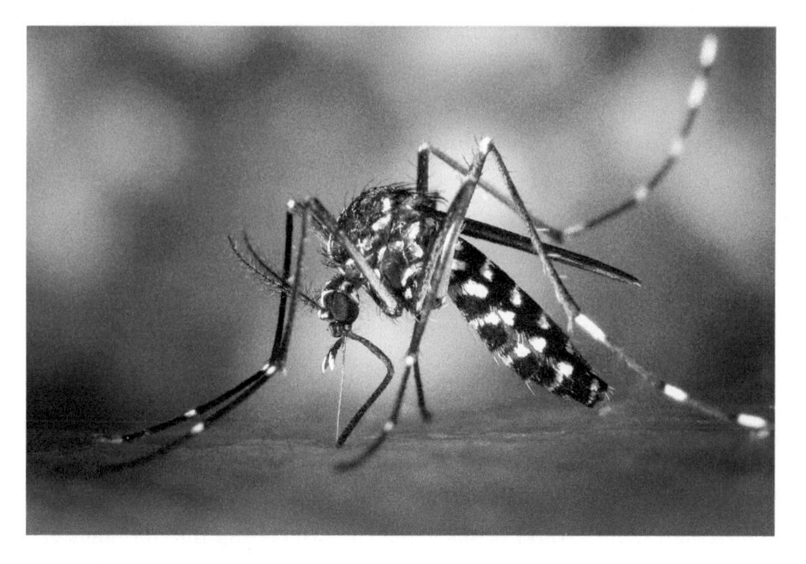

Figure 1.3
A mosquito drills into the skin with her fascicle, a natural microneedle. *Image source: Pxhere.com.*

to understand how they produce a material with such unique properties. Spider silk is a soft, elastic, and biodegradable polymer—yet it has a higher breaking energy than any other natural or synthetic fibrous polymer, far exceeding that of the strongest steel on a weight-for-weight basis. Scientists find inspiration in spider silk—microfluidically extruded through a small orifice in the spinneret called the spigot—to develop new materials for athletics and biomedical applications, among others.[2]

Natural microfluidic phenomena form an integral part of the world within us as well. Every time you inhale—right now—air enters the lung, a wondrous spongelike tree of branching air channels that end in millions of microscopic air sacs (called alveoli). There, blood cells get reloaded with oxygen as in a multitude of microfluidic gas stations operating in parallel. This oxygen exchange is swift and efficient precisely because it happens at a cellular scale, in the small and thin-walled compartment of each blood capillary touching an air-filled alveolus. This microfluidic blood circuit is not unique to the lung. Every cell of your body is maintained by a nearby blood capillary that delivers the necessary nutrients and oxygen and removes the toxic by-products of metabolism.

Perhaps less obvious, but more intimately ours, is the fact that we owe our very existence to microfluidics. During development, all creatures rely on the secretion of gradients of molecules—called morphogens—to attain the final, convoluted shape of the organism. Look at any part of your body, or of any multicellular creature, with a magnifying lens. A hair, a root, a stem, a limb, an organ, could not be formed without intricate networks of tiny distribution channels that bring nutrients and signaling molecules to every cell of the organism. We *are* microfluidic.

MICROFLUIDIC TECHNOLOGY AROUND US

One could write several books about the fascinating microfluidics created by nature. However, this book focuses on microfluidics fabricated by humans. These microfluidic devices surround you on any given day, even though you may not have realized it. Microfluidics are tiny tools for scientists, and they are inside some products you have been using.

Have you heard of inkjet printers? They spit out tiny little droplets of ink with great accuracy. The process is so fast and automated that you do not need to worry about the details, but the droplets are generated by the head of the printer. That printhead is a microfluidic device. These devices are used in every office and many homes and by the industry to produce packaging labels and advertising materials.

If you are one of the nearly eighty million diabetics worldwide that need to take insulin to keep their blood glucose in check, your life may depend on a small tool called a point-of-care glucometer. This microfluidic instrument samples minute amounts of blood using electrodes embedded in channels. However, you do not need to know that there *are* channels, just as you do not need to know that there are transistors in a smartphone to realize how to operate its screen.

And if you are a parent, you might have used a pregnancy test—a microfluidic device based on a paper strip that detects a unique hormone in a woman's urine when she is pregnant. All you need to know is that one line means "no" and two lines mean "yes." The engineers that designed these devices found inspiration in natural microfluidic phenomena, such as the wicking of water by the roots of a tree by a phenomenon called capillarity or the breakup of a garden hose's stream into drops due to a fluid property called surface tension. These are just examples, but there are many others.

Microfluidic engineers, pushed by biotechnology and pharmaceutical companies, struggle to make the devices reliable and easy to use. Customers do not want unreliable, clunky products. Ideally, microfluidic devices should be as foolproof and easy to use as smartphones, although that is not always possible because fluid inlets carry additional complications not present in a smartphone. But the ultimate goal is the same: to conceal the technology so that the user is unaware of its complexity. The operator should only interact with a box and only be concerned with entering information. In a smartphone, the user inputs information by tapping or talking. In a microfluidic device, the user introduces information by adding a fluid or tissue sample. The box "magically" provides the answer—the "sample in, answer out" paradigm. Because of these efforts, you do not see a microfluidic device when engineers successfully develop it into a product. By design, they will always keep it hidden from you. While many research efforts created the core microfluidic technology that you do not see, substantial manufacturing efforts went into safely stowing away the same microfluidic circuits.

Microfluidics is a fertile field that deserves to be inspected from various angles. Microfluidics is both a scientific field and a technological discipline. The mature scientific discipline that examines fluid properties and behaviors is called *fluid mechanics,* so microfluidics may be considered the child of fluid mechanics that studies fluids constrained to submillimeter spaces. In the constricted volumes of a microchannel, flows are eerily devoid of turbulence, and fluid motion can be driven by surface charges or by spontaneous wicking—unlike the turbulent flow coming out of your garden hose or your shower head. These differences between the macroscale and the microscale have elevated microfluidics into a burgeoning engineering specialty, one that exploits the submillimeter confinement of fluids to develop highly efficient, miniaturized assays.[3]

UNDERSTANDING SURFACE TENSION AND CAPILLARY ACTION—THE FIRST MICROFLUIDIC DEVICES

It took some time to appreciate and understand that fluids behave differently at small scales. Isaac Newton's lab assistant, Francis Hauksbee, became a creative experimenter on his own and recorded observations on the wetting of small capillary tubes (figure 1.4)—the first scientific experiments with a human-made microfluidic device. In 1706, he observed that liquids rise higher in tubes of smaller diameters.[4] About a decade later, James Jurin observed that the height of fluid in a capillary tube or column was a function only of the cross-sectional area at the surface, not of any other dimensions of the column (Jurin's law).[5] In the early 1800s,

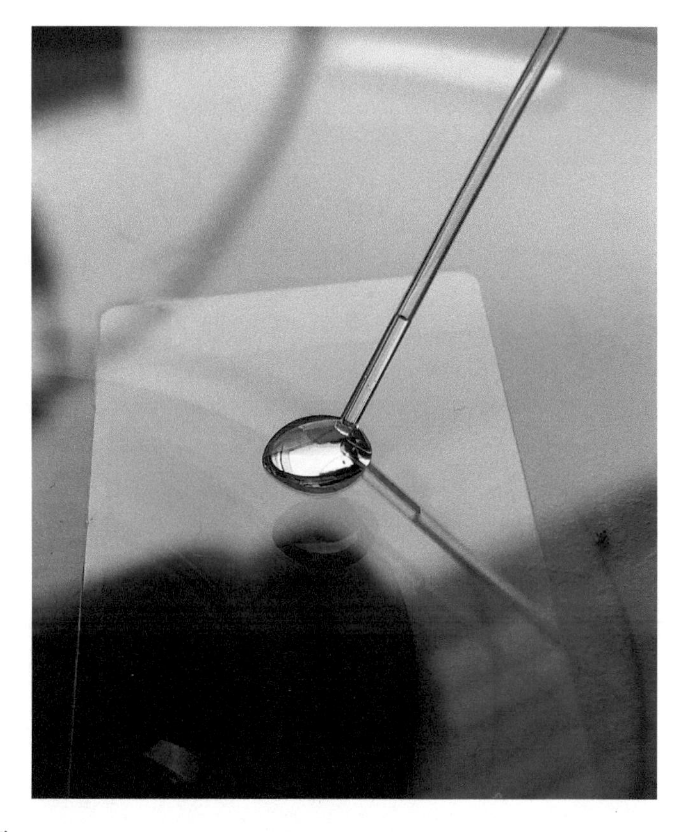

Figure 1.4
The spontaneous rise of water inside a 100-micron-wide glass tube by capillarity. The motion stops at around twice the height shown in the image once balanced by the weight of the water.
Image by Albert Folch, University of Washington.

Thomas Young and Pierre-Simon Laplace gave an expression for the height of the liquid as a function of the contact angle, surface tension, and radius of the liquid column.[6]

To write his results, Hauksbee would have used a quill pen (figure 1.5), a centuries-old microfluidic device based on *capillary action* or *capillarity*—more commonly known as wicking. Wicking is the process by which the energy stored on the liquid surface propels the liquid through narrow

Figure 1.5

Quill pens use capillarity to fill their hollow shaft with ink. The image shows two turkey feathers before and after being made into writing quills. Bottom quill: unmodified. Top quill: cut, hollowed, and polished.

Image by Albert Folch, University of Washington.

spaces. Quill pens are made of feathers from a large bird such as a goose, swan, or turkey. As the tip of the quill is dipped in ink, ink rises into the hollow shaft (called the calamus) of the feather by capillarity. Capillarity is also at work to empty the quill *during* writing. When the writer gently presses the tip of the quill into paper, the ink starts entering the paper's microscopic mesh of cellulose, which—aided by gravity—acts as a capillary pump to pull the ink out of the calamus. The first uses of quill pens date from the sixth century CE, and earlier writing devices such as the reed pens used by Egyptians, Greeks, and Romans to write on papyrus as early as the fourth century BCE also utilized capillarity as the operating principle. As the calamus runs out of ink periodically, quill pens leave an unequivocal, alternating fading trace. Since the manuscripts of Leonardo da Vinci, who lived in the fifteenth century, have a homogeneous trace, we infer that he must have solved this problem by inventing the first reservoir pen, also known as the fountain pen. Modern fountain pens utilize a metal nib with a submillimeter-sized slit that acts as a capillary channel between the reservoir and the paper (figure 1.6).

Surface tension plays a central role in capillarity. You can picture the action of surface tension as an elastic membrane made by water molecules at the surface of water. Where these surface water molecules meet air, they bunch tightly together, causing the elastic behavior. Inside a liquid, molecules are attracted to each other by cohesive forces from all directions. The water molecules on the surface of water, on the other hand, lack water molecules above them. Hence, they hold on tighter to the water molecules on and immediately below the surface. As a result, there is a net inward force that causes water to behave as if a stretched elastic membrane were covering the water surface. All fluids have surface tension, but fluids whose molecules are more cohesively attracted to each other have higher surface tension. Because water molecules have a high attraction for each other via hydrogen bonds, water has a high surface tension—and, for the same reason, a high boiling point. Solvents such as ethanol, on the other hand, have low surface tension and a low boiling point.

Figure 1.6
Modern fountain pens have a metal nib with a slit, which causes continuous capillary wicking of the ink between a reservoir and the paper.
Photo by Aaron Burden on Unsplash.

The "engine" of capillarity is the wetting or *hydrophilicity* of the surface: if the surface is water-loving (such as the glass tubes that Hauksbee used), then hydrophilicity will cause the fluid to crawl up the walls. But as soon as the fluid at the walls of the conduit has crawled up, the elastic surface of the water will be stretched, and it will relax into its stable (flat) position by drawing the column of water up. Next, the edges will wet the walls and crawl up again, successively. The motion only stops when the capillary force equals the weight of the water column. You can now see the utility of a hydrophilic microchannel in a flat chip: no pumps are required to drive flow through it. When the channels are water-hating or *hydrophobic,* capillarity does not work, so filling very small hydrophobic channels with fluids can be challenging.

Even if you have not heard of microfluidics, you must have owned—and still own—a few dozen microfluidic devices that apply the principle of surface tension, such as the ballpoint pen (figure 1.7). The ballpoint pen consists of a rolling metal ball that caps one end of an ink reservoir and is held in place by a socket. The ball can range in diameter from 0.5 to 1.2 millimeters and is made of a hard metal such as steel, brass, or tungsten carbide. The ball plugs the ink reservoir, preventing the ink from spilling over the paper. As the writer scribbles on paper, the rolling action covers the ball in ink. Due to the low surface tension of the ink in contact with the metal surface, only a very thin coating makes it past the opening, but that is enough to deposit ink on the paper.

Figure 1.7
The tip of a BIC ballpoint pen at high magnification. The ink is specially designed with the right surface tension and viscosity to allow for transfer of a thin layer of ink between the reservoir and the outside as the miniature metal sphere rolls when you write. More than 100 billion BIC pens have been sold since 1950, making it the most successful microfluidic device in history. This image is licensed under a CC BY 3.0 license.
Photo by Carlos E Basqueira.

The first commercial realization of ballpoint pens in 1938 was due to the efforts of Hungarian newspaper editor László Bíró, who had become frustrated with the poor reliability of fountain pens, aided by his brother Győrgy, a chemist. Born in Budapest, László Bíró went into exile in Argentina in 1940 during the Nazi occupation. In Argentina—where a ballpoint pen is still called a biro—Bíró started the manufacture of the first ballpoint pens in 1945. The idea was not new, but previous attempts had been unsuccessful at producing reliable pens. Bíró understood that he needed a viscous ink coupled with a ball-socket mechanism that allowed for controlled flow at a very small scale. Unlike previous water-based inks, Bíró's ink was composed of dyes suspended in an oily solvent, such as benzyl alcohol or phenoxyethanol, which form a viscous fluid that dries quickly and helps lubricate the ball tip due to the ink's low surface tension.

Bíró sold the ballpoint pen patent in 1953 to French businessman Marcel Bich, who had a fountain pen company in Paris. At the time, ballpoint pens cost more than $10 (equivalent to about $90 in 2020). The company (now called BIC) had reengineered Bíró's ballpoint pen to lower the manufacturing costs and had started mass producing the ubiquitous BIC Cristal ballpoint pen in 1950 under a license from Biró. BIC sells about five billion pens a year and has already sold more than 100 billion pens since 1950. It is found in all corners of the world. By comparison, everyday items like smartphones and cars are a rarity: there are 3.5 billion smartphones, 2 billion computers (counting laptops, desktops, and servers), and 1 billion cars in the world today, and we cannot say yet that most humans own a phone, a computer, or a car. On the other hand, thanks to BIC, we can affirm that most humans have used a microfluidic device.

Surface tension is also responsible for holding water drops together (figure 1.8). There are raindrops, dewdrops, mist. There are also less obvious phenomena that can only be observed with high-speed cameras. Ultrafast photography has allowed us to capture the breakup and splash of drops on a pool of water, forming symmetrical liquid crowns whose existence is as brief as they are mesmerizing (figure 1.9). Belgian physicist and mathematician Joseph Plateau observed in 1873 that liquid streams break into

Figure 1.8
Drops on the petals of a daisy flower. The drop shape is held by surface tension. *Image source: Pxhere.com.*

droplet trains that are too fast to be visible to the naked eye (figure 1.10). This breakup is more obvious as you gradually slow down the fluid flow of a faucet to a drip, when you can observe the breakup of the droplets near the mouth of the faucet, but is not evident as you are watching a display fountain. Why are liquid streams unstable? Lord Rayleigh provided a straightforward explanation a few years after Plateau's observation: as soon as the water exits the pipe and stops feeling the wetting contact of the pipe wall, the surface tension of water is so high that it starts slicing the stream like a loaf of bread. As the water tries to minimize its surface energy, it forms the most stable of shapes: spheres. This simple observation is the basis of the field of *droplet microfluidics* and many microfluidic inventions, such as the inkjet printer (chapter 2).

Figure 1.9
Ultra-high-speed photography of a water drop as it bounces off a water surface. *Image source: Pxhere.com.*

UNDERSTANDING DIFFUSION AND LAMINAR FLOW

Scientists have long known that fluids confined to small volumes have intriguing properties. Until microfluidics became widespread, chemical assays were cumbersome and slow and wasted a lot of reagents. A chemical reaction typically required filling a Pasteur pipette and mixing its contents with those of a flask by manual stirring. The typical quantities of fluid involved were on the order of a fraction of a milliliter at best and reaction times were on the order of at least seconds. As smaller and smaller volumes are required, a mixing mechanism other than stirring becomes more and more relevant: *diffusion*. Diffusion results from the natural tendency of molecules to bounce against each other, and, in doing so, spread far from their initial position. Many microfluidic devices exploit the fact that, in small volumes, diffusion acts very quickly so stirring is not necessary. In a small volume like a drop of water, molecules

Figure 1.10
High-speed photography of a water jet reveals the breakup of the stream into droplets. This breakup is imperceptible to the naked eye.
Image by Albert Folch, University of Washington.

in their random motions quickly find the end of the drop and exchange position with other molecules or "mix."

One of the great wonders of microscopic flow in a microchannel is how predictable it is (see box 1.1). Like in a river, fluid flows faster at the center and very slowly at the walls. For a cylindrical channel, the mathematical function that describes the flow speed distribution is an exact parabola; hence, we speak of a *parabolic flow profile.* Because all the flow lines are

BOX 1.1
Microfluidics Terminology—Flow

Microfluidics can be seen as the child of fluid mechanics, and, as such, microfluidics has inherited a wealth of knowledge from its progenitor. Leonhard Euler (1707–1783) was a Swiss mathematician that first derived the complete equations for the flow of idealized fluids that lack friction and therefore have no viscosity. However, as all real fluids are essentially viscous, the predictions of Euler's equations did not accurately predict the experimental outcomes. For a while, scientists thought that hydrodynamics was an impractical theory. French mechanical engineer Louis Marie Henri Navier (1785–1836) first derived the equation for the movement of a viscous fluid by applying Newton's second law of motion ($F = m \times a$) to an elementary rectangle of fluid. In 1845, British physicist George Gabriel Stokes (1819–1903) perfected the derivation of the now-called *Navier-Stokes equation*. Unfortunately, the equation has terms that render it nonlinear. A few special flow profiles could be solved analytically: the laminar flow between parallel plates, flow in a round tube, and flow in a rectangular-cross-section channel.[7]

To attack this difficult problem, some resorted to experimentation. French physician Jean Louis Poiseuille (1799–1869) studied the pumping power of the heart and the movement of blood in vessels and capillaries. His experimental equation describes that the flow rate in a small glass capillary is proportional to the pressure differential and to the fourth power of the diameter and inversely proportional to the tube length—known as the *Hagen-Poiseuille law*, which students now derive as one of the solvable cases of the Navier-Stokes equation. To study complex states of flow, physicist Osborne Reynolds (1842–1912) used a jet of dyed water centered within the flow of another, larger pipe. In 1883, he discovered that a laminar flow turns to a turbulent flow when, independently of the values of the average velocity v, glass tube diameter d, water density ρ, and water viscosity μ, the value of the nondimensional quantity $\rho v d/\mu$—now named *Reynolds number*—reaches the value of ~2300–2900. The Reynolds number, as it turns out, equals the ratio of the inertial force and the viscous force. Several other dimensionless numbers have been introduced subsequently to characterize various flow properties.[8] They have become a useful tool for microfluidic engineers to design systems even when the Navier-Stokes equation cannot be solved.

With modern computers, it is now possible to obtain *numerical solutions* of the Navier-Stokes equation (i.e., software packages that use finite-element methods) for almost any given microchannel geometry. But the first microfluidic engineers did not have computers as fast as the present ones and could not easily fabricate any of the three geometries for which analytical solutions to the Navier-Stokes equation existed (not even the rectangular profile) until the late 1990s, so flow modeling was challenging for a few years.

parallel, this profile facilitates another characteristic of microscopic flow: *laminar flow* (figure 1.11). Laminar flow is a parallel mode of mass transport that occurs in the absence of turbulence and that sometimes seems to defy the laws of diffusion. The flow is laminar when *viscosity*, a property of a fluid that resists flow, becomes more important than *inertia*, the tendency to remain in its state of motion. Several fluids flowing in a microchannel appear to flow side by side, as if not mixing, although in reality what we see is a snapshot of the diffusion process as the fluids flow downstream. On the other hand, if inertia overcomes viscosity, the flow becomes turbulent. We can avoid turbulence using a thick, viscous fluid like honey, or by using slow fluid flow velocities—or by constraining the flow in microchannels.

In considering these phenomena, scientists look at *scaling laws* because physical objects have relationships that are universal: when the dimensions of a sphere or a polyhedron shrink, its volume shrinks as the cube of the width, whereas its surface area only shrinks as the square of the width. For example, if you reduce the size by two, the volume will shrink by eight, but the surface area will only shrink by four; if we reduce the size by three, the volume will shrink by twenty-seven, but the surface area will only shrink by nine; and so on. In other words, the *surface-to-volume ratio* grows as an object shrinks. In the large surface-to-volume ratios of a microchannel or of a droplet, the borders are very close to the center, effectively constricting the fluid motion and increasing the apparent viscosity.

If you want to imagine what laminar flow feels like, picture yourself as a tiny swimmer. Water feels to a moving bacterium—an entity with a tiny mass and therefore negligible inertia—as swimming through a thick, viscous liquid like honey or molasses would feel to you. In a microchannel, flow and diffusion become two intimately connected modes of mass transport (see box 1.2). You can try to picture the relative contribution of flow and diffusion to mass transport: if molecules were people, mass transport by flow would be the transport of people by cars—allowing people to go fast but constrained to specific directions—whereas mass transport by diffusion would be the pedestrians—allowing people to go in

Figure 1.11

Sixteen dye streams seem to not mix when they are squeezed into one channel as flow goes from left to right. The brief amount of time the dyes spend in contact with each other as they travel from the left of the image to the right does not result in sufficient dye diffusion for the eye to detect. Although the interface between the blue and the red dyes suggests that they are diffusively mixing into each other, we know from the other dyes that they have not had time to mix yet. The black lines are mostly an optical effect due to the slight inclination of the fluid interface arising from a channel wall slant, a fabrication defect. What we see as dark bands was created by light going through both the red and blue dyes as they flow in (slanted) parallel sheets or laminar flow, but the dark bands disappear as the streams are squeezed into a vertical orientation between two parallel, vertical walls.

Image courtesy of Greg Cooksey and Albert Folch, University of Washington.

arbitrary directions, but slow. In microfluidics, the "streets" are narrower, with all the cars being forced to circulate in parallel, so changing lanes is difficult. Thus, microfluidic engineers have incorporated laminar flow into their designs in rich and imaginative ways to control the diffusive spread of chemical compounds.

Electrophoresis—the separation of mixtures using electrical fields—has played an important role in the history of microfluidics. Researchers

BOX 1.2
Microfluidics Terminology—Diffusion and Heat

The first systematic study of diffusion (in gases) was performed by Thomas Graham in 1831–1833, who noted that "gases of different nature [. . .] spontaneously diffuse."[9] Inspired by Graham's findings, the young German physiologist Adolf Fick in 1855 formulated a set of partial differential equations—Fick's laws of diffusion—that predict how diffusion causes the concentration of a solute in a liquid to change with time. From these equations, we find that the time it takes for a molecule to diffuse over a certain distance is proportional to the square of the distance. In other words, it takes a molecule four times longer to diffuse twice as far. Or, as the saying goes in microfluidics, "10× smaller is 100× faster."

Diffusion and flow also affect heat transport: heat is easier to remove for thinner or smaller samples due to the higher surface-to-volume ratio. The law of heat conduction, or Fourier's law, formulated by French mathematician Jean-Baptiste Joseph Fourier a couple of decades earlier, is formally identical to Fick's second law of diffusion. In particular, the separation of mixtures using electrical fields, dubbed *electrophoresis*, causes unwanted heating of the fluids, so minimization of the volume requires smaller voltages and minimizes heat production. Hence, in 1940, J. St. L. Philpot published an electrophoresis device based on continuous flow in a thin sheet of fluid confined between two plates.[10] In 1948, Gotfred Haugaard and Thomas D. Kroner of the United Shoe Machinery Co. pioneered the idea of using a piece of filter paper as an inexpensive medium that both restricts and generates flow within the plane of its porous matrix by wicking. They reported the separation of amino acids with applied voltage (~100 volts) using paper,[11] which was widely used thereafter. A generalization of the process was presented in 1950 by W. Grassman and K. Hannig.[12] These pioneering studies concluded that a further reduction in fluid confinement could lead to further improvements.

realized early on that reducing the fluid volume would reduce the required voltage and would also result in reduced heat production, an undesired effect of running a current through the fluid. In 1981, J. Jorgenson and K. Lukacs demonstrated electrophoresis in glass capillaries—that is, cylindrical microfluidic channels—to reduce the separation time and heat generation.[13] The reduced heating observed in *capillary electrophoresis* is due to the large surface-to-volume ratio of the capillary channel, the thin capillary walls (50–150 microns), the high thermal conductivity of the wall material, and the low current passing through the capillary. The improvements afforded by miniaturization in capillary electrophoresis were noticed by NASA scientists, who considered sending micromachined electrophoresis chips to space.[14] We do not have any record of any of these researchers using the term *microfluidic* at the time.

SILICON MICROTECHNOLOGY

It is not by chance that microfluidic chips were born from microelectronics, one of the most transformative collective engineering developments of our species. Hence, the same microfabrication technology that underlies microelectronics later led to the first microfluidic devices.

Computer size and performance depend on the size of its unit components. In the 1940s, engineers built electronic computers with vacuum tubes, but those early computers occupied a large room, generated enormous amounts of heat, and were unreliable. In 1947, American physicists John Bardeen and Walter Brattain, working under William Shockley at Bell Labs, invented an ingenious device called the *transistor* for which they received the 1956 Nobel Prize for Physics. They made it with germanium, which is also a semiconductor and the closest chemical relative of silicon. They could use transistors to amplify electrical power or switch electronic signals. The transistor quickly replaced the vacuum tube in computer architecture, so by the late 1950s and 1960s, computers featured circuit boards filled with hundreds of individually soldered transistors. Thanks to the transistor, the size of computers shrunk by an order of magnitude—to as big as a large table by the early 1960s. However, the transistors had to be soldered one by one, so scaling up computers to the millions of transistors needed for more complex tasks was a daunting prospect.

Then the *MOSFET* was invented. In 1959, Mohamed Atalla and Dawon Kahng at Bell Labs realized that germanium could be substituted by silicon to fabricate a planar and much more compact transistor. Planar transistors could be miniaturized and mass produced in a clean room (figure 1.12) using photographic projection principles. Because of the architecture and operating principle of the device, they named it a metal–oxide–semiconductor field-effect transistor, or MOSFET.

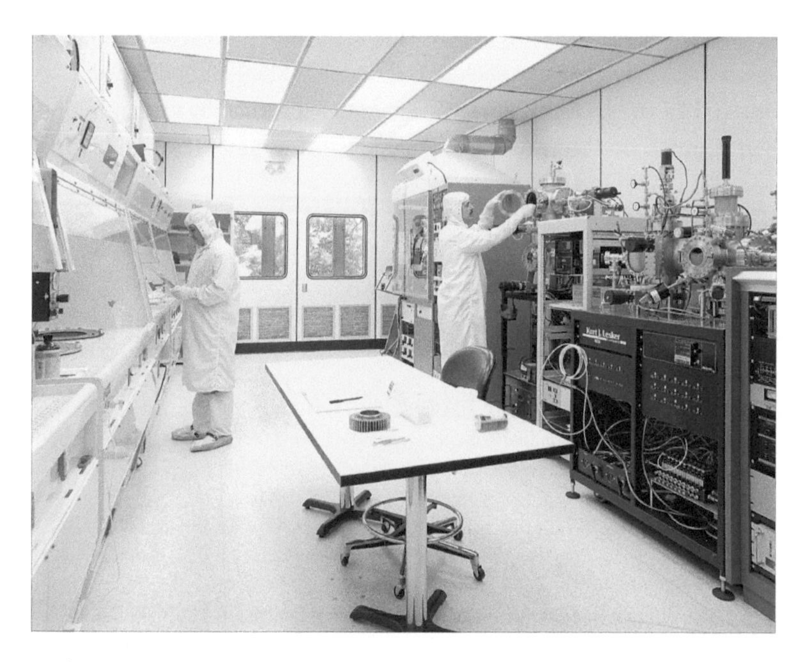

Figure 1.12
A microfabrication clean room. The air is filtered and engineers must fully gown in bunny suits to avoid dust contamination. Yellow light can be used because the photoresists are not sensitive to it. This file is in the public domain in the United States because it was solely created by NASA.

In the early 1970s, a few pioneers wondered if the photolithography techniques for making MOSFETs (see box 1.3) could be used for making miniature electromechanical components. These researchers started building the first *microelectromechanical systems* (*MEMS*) in silicon wafers. Japan had a vibrant microelectronics program, so many MEMS pioneers were from Japan. Researchers fabricated MEMS devices with sequences of photolithography, chemical etching, and deposition steps applied to a silicon or glass wafer,[15] resulting in planar devices on the surface of the wafer. A famous example of MEMS is a miniaturized accelerometer that is now ubiquitously used as an airbag deployer.

Making smaller components not only reduced device footprint and weight, but it also improved performance. Among these early MEMS researchers, a few pioneers such as Jim Angell at Stanford and Ernest Bassous and Lawrence Kuhn at IBM wanted to show that, by miniaturizing fluid channels in silicon, one could manipulate tiny amounts of fluids and

BOX 1.3
The Success of the MOSFET Transistor

To fabricate a MOSFET, Atalla and Kahng at Bell Labs used a technique called *photolithography*. The process of photolithographic patterning requires several steps. First, one covers a mirror-polished silicon substrate (called a *wafer*) with a thin layer of a photosensitive polymer (termed *photoresist*). Next, one selectively exposes the photoresist to light of the appropriate wavelength (typically ultraviolet), either employing a laser scanned across the surface or through a *photomask* (a plate with opaque and transparent features that define the design of the circuit, figure 1.13). The photomask can also be projected to obtain the benefit of optical reduction, or exposed with high-resolution lasers. After exposure, one develops the photoresist layer in a particular developer solution, much like in photographic development. This step exposes or protects the underlying silicon areas according to the pattern. Next, one deposits other materials (e.g., metal) or etches the substrate in the areas not covered by photoresist. By repeating this process several times, it is possible to build complex thin-film circuits with more than sixty micropatterned layers and fabricate several circuits at once on the same wafer (*parallel fabrication*).

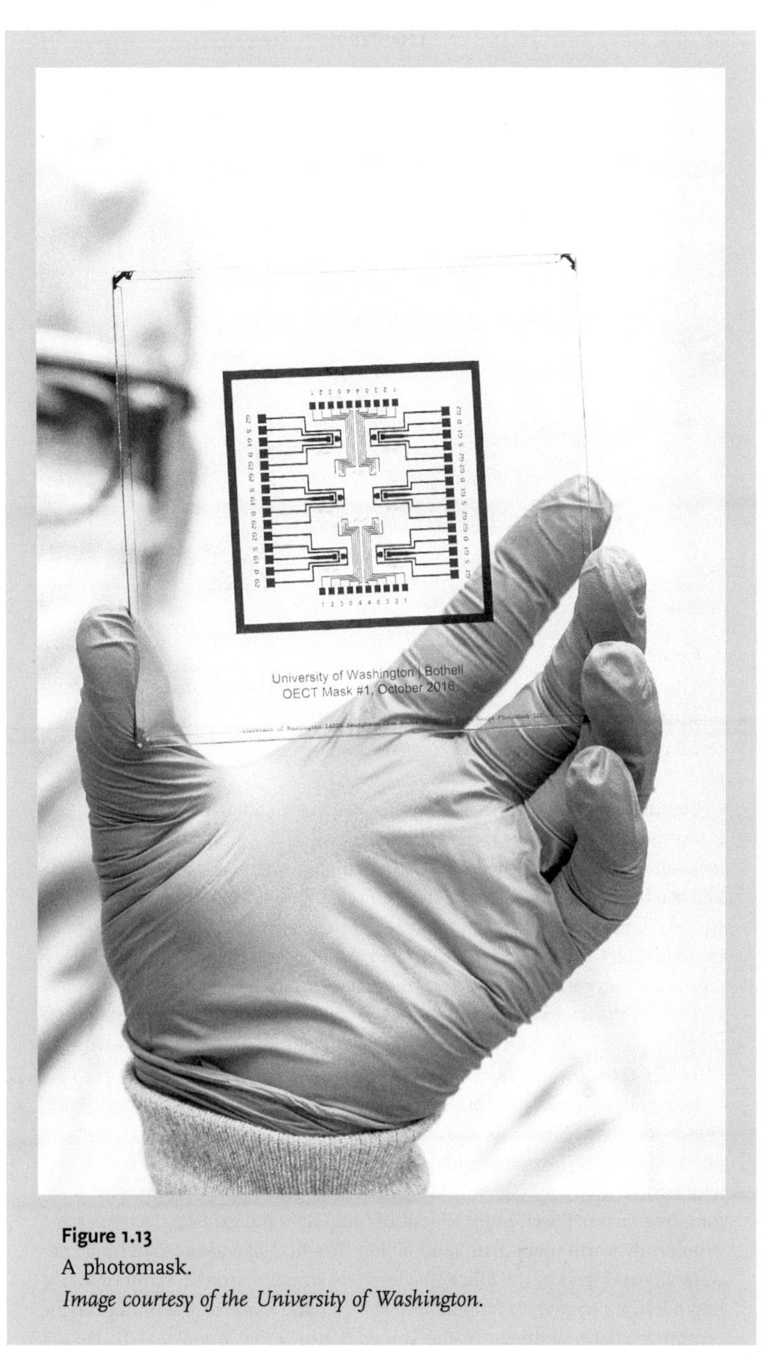

Figure 1.13
A photomask.
Image courtesy of the University of Washington.

(continued)

BOX 1.3 (continued)

Figure 1.14
The motherboard of a modern computer. Each black integrated circuit square contains millions of MOSFET transistors.
Image by Albert Folch, University of Washington.

Because optical projection works well down to the submicron scale, the industry has been using the approach since the 1960s to pack more and more transistors and other circuit elements into a wafer. Engineers can automate the process and batch process wafers to obtain the benefit of low-cost manufacturing that is typical of microelectronics. Electronic engineers use a modular approach to build more complex systems: transistors form the building blocks of integrated circuits, microprocessors (figure 1.14), and microcomputers. A typical smartphone consists of hundreds of billions of transistors. It is estimated that thirteen sextillion (1.3×10^{22}) MOSFETs have been manufactured since 1960, making the MOSFET the most widely manufactured device in history.

Silicon-based chips have been an enormous source of innovation and wealth worldwide. In 1968, chemist Gordon E. Moore (now known for Moore's law) and physicist Robert Noyce, coinventor of the integrated circuit, founded a company, later named Intel, in Mountain View, California, just 5 miles southeast of Stanford. Intel makes the chips for billions of computers and gaming electronics and is presently worth more than $200 billion. The first silicon-based technologies were all developed in the Silicon Valley area, a region around Stanford that is now a leading ecosystem for high-tech startups and home to one-third of all the venture capital investment in the United States. After World War II, Japan's economy was rebuilt in record time by, among other reasons, concentrating on the emerging industry of chip manufacturing (e.g., Fujitsu, Hitachi, Mitsubishi, NEC, and Toshiba, the Big Five Japanese chip makers). Japan's economic resurgence became an example for other countries in East Asia such as South Korea, Taiwan, and China, which also invested heavily in developing a booming consumer electronics and chip manufacturing industry.

save reagents and time. They did not know it then, but they were starting a new era in science and technology—the era of microfluidics.

SILICON AND GLASS, THE FIRST MATERIALS FOR MICROFLUIDICS

Silicon, the basic MEMS material, was the first material at the disposal of researchers to fabricate microfluidic channels (box 1.4). Regrettably, it is hard as a rock, and nature has not given us many options for etching silicon. A wafer left in a bubbling hot bath of potassium hydroxide (KOH) dissolves in a couple of hours. While KOH etching is relatively simple, the manufacture of deeper or more vertical etches required more sophisticated equipment and chemistries such as highly ionized gases (called plasmas). The plasma is activated by energizing a coil with a high-frequency, high-voltage electrical field within a chamber that is exhausted to a safe pipe and neutralization conduit. Unfortunately, the gases whose plasmas are very reactive with silicon are also highly toxic. They are so dangerous that, in clean room facilities, any leaks will cause the evacuation of the whole building. The laboratory managers usually keep the tanks in a separate room full of sensors just in case. An additional problem with silicon is that it is opaque, so it was difficult to visualize what was going on inside the channels during those first experiments.

An obvious alternative is glass, which is silicon dioxide—a material with which a MEMS engineer is familiar because the surface of any silicon wafer becomes oxidized in contact with oxygen or air and covered with a layer of silicon dioxide. Glass also offers many advantageous qualities for

BOX 1.4
Overview of the Materials Used in Microfluidics

Microfluidic devices have many applications, and each application has different material requirements. Therefore, microfluidic engineers have devised different techniques to build microfluidic devices with various materials, such as polymers, glass, or paper. Unfortunately, it is still challenging to build microfluidic devices in more than one material. We teach our students that the material chosen to make a microfluidic device tends to determine the range of applications that we can use the device for—so choose wisely. We can fabricate microfluidics in three broad classes of materials.

The materials in the first class are hard and brittle: *silicon and glass*. Because microfluidic technology was born from microelectronics, the first microfluidic chips were made in silicon wafers (chapter 2). Silicon, however, is not ideal because it is not transparent. The transition to glass was easy: silicon is chemically compatible with glass since silicon is immediately oxidized on contact with air to form silicon dioxide (which is essentially glass). Glass is the perfect material for chemical applications for the same reasons a chemist uses it for test tubes and running DNA gels: it is inert and transparent and its surface is brimming with negative charges. These surface charges are beneficial for initiating chemical reactions. They are also essential for controlling flow using voltages (chapter 3). Yet both silicon and glass are inert materials that require expensive etching processes in dedicated processing facilities. Also, in the early 1990s, the devices were limited to thin, ~500-micron-thick wafers. It would take almost a decade before the third dimension would be conquered by stacking laminates or by 3D printing.

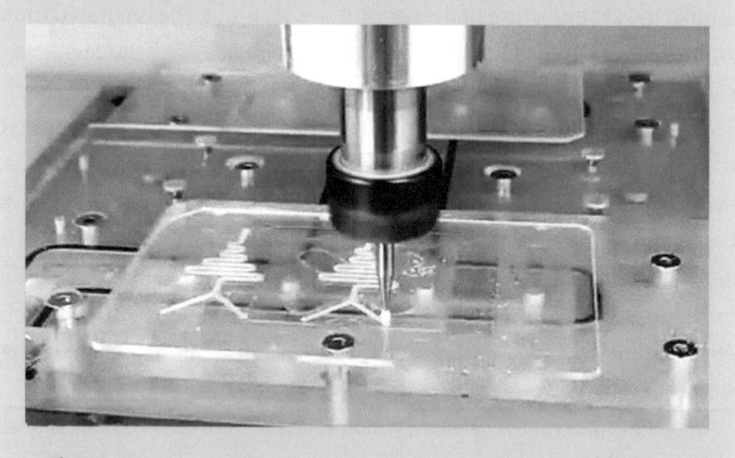

Figure 1.15
Many thermoplastics are good materials for microfluidics because they are transparent and biocompatible. They can also be easily micromilled because they are soft.
Image courtesy of Steve Soper, Kansas University.

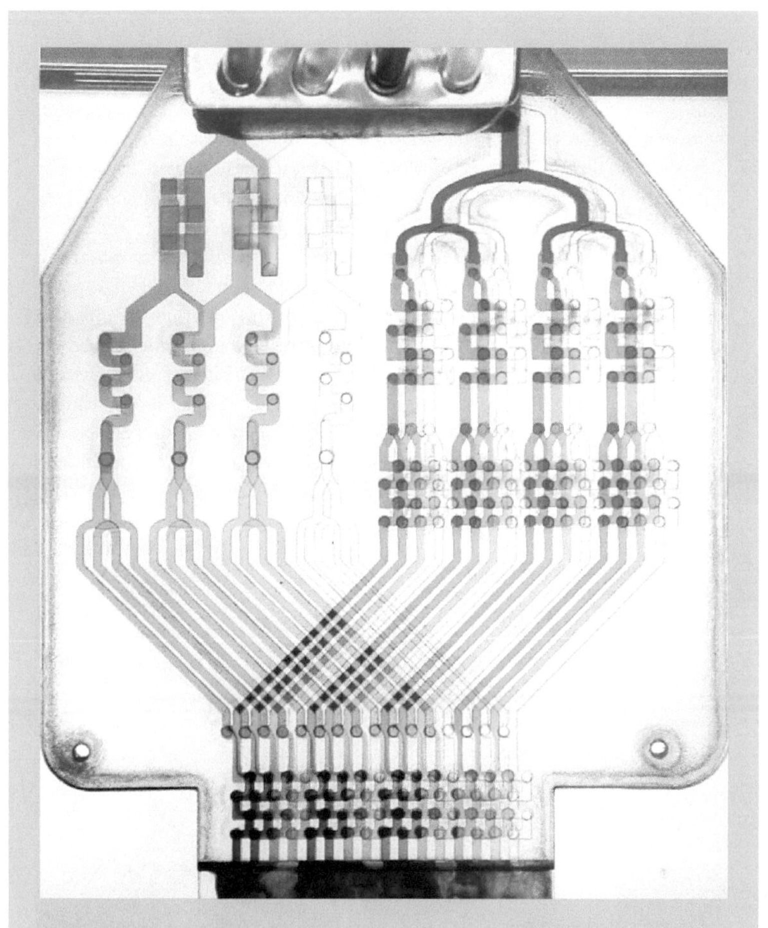

Figure 1.16

Thermoplastics can also be laser cut with high precision. This image shows a combinatorial mixer where fluids are flowing at four different levels. The mixer has been fabricated courtesy of laser cutting and stacking of nine mylar laminates.

Image courtesy of Chris Neils and Albert Folch, University of Washington.

The materials in the second class are soft and flexible: *polymers*. Unlike silicon and glass, chemists can inexpensively synthesize large quantities of polymers. Researchers can easily machine (figure 1.15), laser cut (figure 1.16), or mold polymers (figure 1.17) into various shapes and stack them to build channels. Microfluidic engineers use three subclasses of polymers. *Thermoset polymers* irreversibly "cure" into the final solid polymer when the temperature is raised, much like egg white changes into a white substance when it is heated. *Thermoplastic polymers*

(continued)

BOX 1.4 (continued)

can be reversibly melted back and forth above and below their glass transition temperature, much like butter. *Photopolymers* (short for photosensitive polymers) are polymers that form or break down via a chemical reaction that is triggered by light (called a photochemical reaction), like the photoresists used in photolithography (box 1.3) or the resins used in 3D printers (chapter 9).

The most widely used thermoset polymer in microfluidics has been *poly(dimethylsiloxane)* (*PDMS*). PDMS is a clear, medical-grade, rubberlike material (figure 1.18). Typically, researchers make the PDMS polymer by mixing a prepolymer and a curing agent and pouring the mixture atop a microfabricated master; after curing, the solidified PDMS is separated from the mold by hand (figure 1.17). PDMS is transparent, elastic, gas permeable, and biocompatible, making it ideal for microvalves (chapter 5) and organ-on-a-chip applications (chapter 6). Its glass-like transparency has enabled a variety of optofluidic devices that combine microfluidic and optical sensing capabilities (figure 1.19).

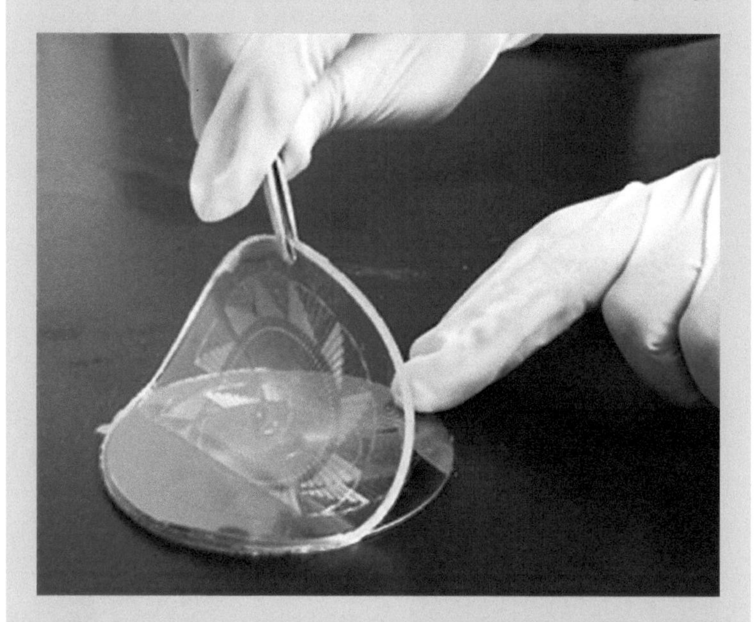

Figure 1.17

In academic settings, most microfluidic devices are fabricated by PDMS molding. The image shows the separation of a PDMS replica from a microfabricated mold on a silicon wafer.

Image courtesy of Gang Li, Shanghai Institute of Microsystem and Information Technology, Chinese Academy of Sciences, Shanghai, China.

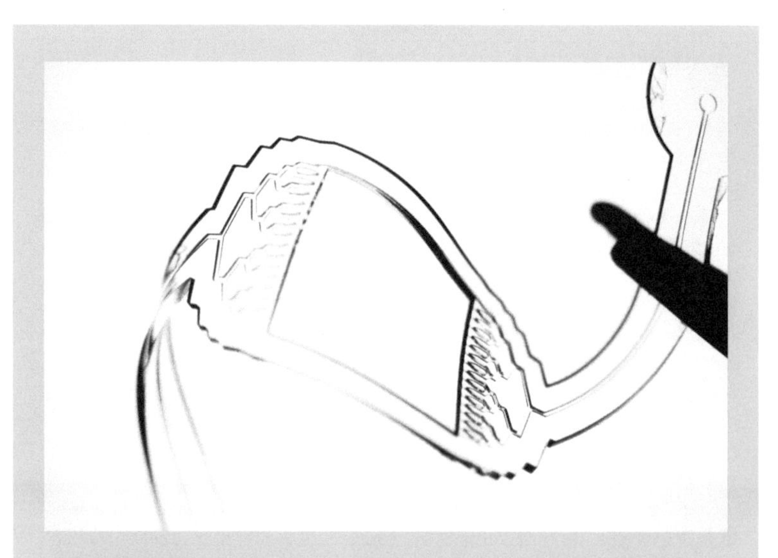

Figure 1.18
A 1-millimeter-thick microfluidic device made in PDMS is being held by tweezers to illustrate the flexibility of PDMS.
Image courtesy of Chris Sip and Albert Folch, University of Washington.

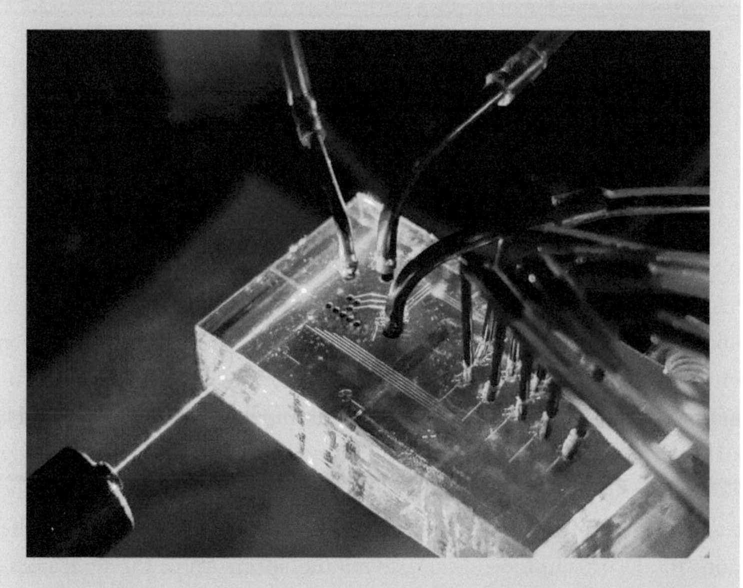

Figure 1.19
A PDMS microfluidic device that uses the light of a blue laser (bottom left in the image) for DNA detection.
Image courtesy of Holger Schmidt, UC Santa Cruz.

(continued)

BOX 1.4 (continued)

Industrial manufacturers use thermoplastics to fabricate most of the plastic objects that surround us in our daily life. Examples of thermoplastic polymers used in microfluidics are polystyrene, poly(methyl methacrylate) (PMMA, also known as Plexiglas), and polycarbonate. These polymers are typically molded using heat. Hot embossing is a technique for molding microfluidic devices at low throughput using a hot metal mold pressed against a flat sheet of the polymer. Injection molding is a technique based on flowing molten plastic into a cavity mold; microfluidic engineers use it for molding parts at high throughput. These techniques were applied to microfluidics in the late 1990s and have proven very useful for commercializing point-of-care devices (chapter 4).

The materials in the third and last class are porous, such as *paper* (chapter 7) or *hydrogels* (chapter 9). Hydrogels are hydrophilic (water-liking) polymers that form a three-dimensional network where the voids or pores are filled with water. Researchers have built microchannels in hydrogels to imitate biological capillaries. It may seem counterintuitive to use a material that lets water through to run fluids inside. Still, most biological materials are porous precisely to allow the transport of molecules (including water) essential to life. Water, buffer ions, or other biomolecules fill the pores in a hydrogel, constituting an ideal environment for cells: hence, hydrogel microchannels can offer unmatched advantages of biocompatibility. The porosity of dry paper confers paper its wicking property and conveniently allows for building plantlike capillary pumps that require no external power and are inexpensive and portable. Indeed, nature offers many lessons for the design of microfluidic devices.

microfluidic devices that would be obvious to a biologist and a chemist. First, it is the same *transparent* material that microscope slides and test tubes are made of. It sustains cell growth, that is, it is *biocompatible*. For centuries, chemists sought the help of glassblowers to make glass instruments with sophisticated shapes such as pipettes, droppers, funnels, beakers, graduated cylinders, and distilling columns. Glass is extremely *inert*, so it can hold a wide variety of chemical reactions. Conveniently, it allows for optical inspection of the reactants, either by eye or by more modern techniques that analyze the spectrum of the light shining through a material—termed *spectrophotometry*. Also, glass—like silicon—has a well-defined surface chemical composition with abundant negative charges. These negative surface charges are useful for anchoring particular classes of molecules using *surface chemistry*, for wetting reproducibly, and for driving and controlling flow inside glass capillaries using voltages (*electroosmosis*). Electroosmotic flow has been widely used in microfluidics. Electroosmosis is the motion of liquid induced by an applied voltage; when an electric field is applied to the fluid (usually via electrodes placed at inlets and outlets), the net charge adjacent to the wall surface is induced to move by the resulting electric field. Many microfluidic engineers vouch that glass has no real substitute.

But the sublime chemical inertness of glass also makes it exceedingly challenging to fabricate microchannels in it. From today's perspective, the research efforts of the first microfluidic engineers who worked with

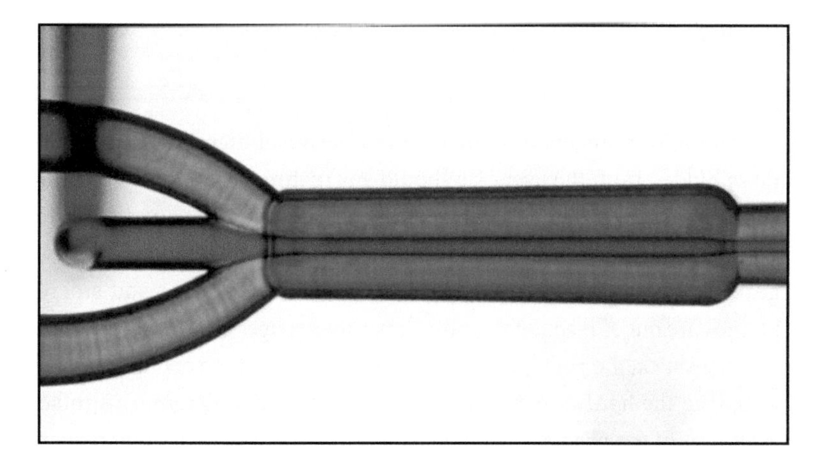

Figure 1.20
A glass microfluidic device that has been fabricated by femtosecond-laser micromachining followed by chemical etching by HF. The laser was focused below the surface and only modified the inside of the glass, so no bonding was necessary. In the image, the device was used to demonstrate the hydrodynamic focusing of a red dye stream with two blue dye streams.
Image courtesy of Petra Paiè, Francesca Bragheri, and Roberto Osellame, CNR—Istituto di Fotonica e Nanotecnologie (Milano, Italy).

glass or silicon look heroic. An important limitation of silicon and glass microfabrication is that researchers cannot do it in small laboratories. It requires the use of extensive, centralized clean room facilities. Chemists developed a few wet and gaseous etches for silicon, but these etches required significant training and infrastructure. There are even fewer options for etching glass: the only known etchants are hydrofluoric acid (known as HF, see figure 1.20) and a mixture of HF and nitric acid, both so hazardous that any contact with the skin, any minor splash, or suspicion of it, warrants immediate hospitalization. These challenges required that administrators centralize the resources in a shared, federally funded facility, and only a few selected centers could afford a clean room.

To make a political analogy: before the 1990s, access to microfluidic fabrication was not democratic. Glass and silicon were antidemocratic and new microfabrication processes based on polymers that allowed for low-cost replica molding were needed (chapters 4 and 5).

* * *

Microfluidic engineers have ingeniously designed their devices to keep them hidden from the user—as should any mature technology: while the first cars trotted about with exposed engines and the first computers had their electronic guts uncovered for the user to tinker with, nowadays the gears of cars or the microelectronic circuits of our laptops and smartphones are out of reach of the nosy consumer: a user interface—the shift or the joystick, the pedals or touchscreen—passes on the operator's commands to the hardware. Microfluidic devices have undergone a similar evolution in the shadows of public attention, but it is now time to open the hood of these devices and tell you the stories of how they came to be.

2 THE POWER OF DROPLETS

Tiny water drops surround us everywhere. Since the beginning of time, nature has been creating droplets, each liquid sphere a testimony of how surface tension can package molecules into a minute pearl-like container. Clouds are formed by a mixture of microscopic water droplets and ice particles, so small and light—10 microns in diameter each on average—they can be carried by the wind. A cubic meter of cloud, fog, or mist contains several hundred million droplets, each separated from its nearest neighbor by about 1 millimeter. When light hits the cloud or the fog, the rays are deflected by each droplet in all directions, giving it the appearance of being white even though each droplet is perfectly transparent.

Then there are raindrops, dewdrops, and, of course, our own sweat drops, tears, and the contentious spitting droplets against which we are asked to wear masks to stop the spread of COVID-19. If they were not so deadly, you would have to marvel at these little entities we call coronaviruses that have evolved to cozily travel through the air in minuscule water spheres. As you exhale, spit, or cough, you are launching viruses like nano-astronauts on an interbody trip in their droplet shuttles, ready to colonize the body that inhales them.

But many of the droplets that exist around us in our everyday lives have been intentionally manufactured by humans. Oily droplets are packaged in a container for your benefit or enjoyment. The salad dressings that you eat every day contain millions of them in the form of emulsions. You might have spread them on your skin when you applied moisturizer this morning. Droplets also have many applications in science and technology. Hence, there have been many efforts at designing devices that generate droplets in large quantities from a liquid stream. Car manufacturers and graffiti artists use sprays that project paint droplets to produce even coats of color. If you have cleaned a window recently, chances are that you sprayed a mist of cleaning solution on the glass surface. If you have asthma, you probably use a nebulizer or an inhaler, devices that diffuse millions of droplets of a bronchodilator such as albuterol deep into your throat to open your airways. Or if you are reading this on a hot summer night, you might be cooling yourself with a spray bottle on its mist setting (figure 2.1). These devices are

Figure 2.1
A spray bottle is used to create a mist, a suspension of fluid droplets in air. *Image source: Pxhere.com.*

microfluidic machines that, inspired by Plateau-Rayleigh instability, project a gas stream against a fluid stream to accelerate the instability and quickly break down the fluid stream into a bouquet of aerial droplets.

A widespread application of airborne droplets is in digital printing. You have likely used one or more types of *inkjet printers* in your office or at home. Many of the next wave of printers—3D printers—are also equipped with inkjet nozzles. The nozzles deposit layer after layer of photopolymers that are rapidly cured with ultraviolet light. Inkjet printers were some of the first microfluidic devices ever invented. But before modern inkjet printers and microfluidic chips came into existence, another form of printing based on continuous inkjets was the mother of them all.

CONTINUOUS INKJET PRINTING

The development of inkjet printing has its roots in early studies on the behavior of fluid jets. Almost seventy years after Lord Rayleigh explained the fluid jet breakup into droplets in 1873, a Swedish engineer named Rune Elmqvist, working for the company that later became Siemens, picked up the Plateau-Rayleigh instability idea and turned it into a practical device. Elmqvist proudly appears in old photographs with hair combed back and dressed in an impeccable suit and tie next to his inventions. A doctor and a brilliant engineer, he became better known later for inventing the first implantable pacemaker. In 1948, Elmqvist wanted to build an automated printing machine that could be connected to one of the new computing Siemens machines to print traces like those of an electrocardiogram. Computers at the time had mechanical printer rolls and punch cards, but these were annoyingly slow. From Lord Rayleigh's studies, Elmqvist knew that, if he created liquid streams of aqueous ink, the jets of ink would break down into droplets. Thanks to the electrostatic charges acquired by the droplets in their friction with air, he realized that electrical fields generated by magnetic coils would deflect the electrified droplets like ink bullets. He described his brainchild in a patent in 1951.[1] The device included a reservoir and a pump for recycling the continuously flowing ink that had to be deflected away from the paper. Siemens used this invention to build the Mingograph, an inkjet-based analog medical chart recorder. Robert Sweet published a similar continuous-printing concept in 1965,[2] leading to the highly successful, still-active printing company VideoJet Technologies (starting as a division of A. B. Dick)—likely the first microfluidics-based company in history. At the time, inkjet technology development was intense, with IBM and others as foremost contenders.

Continuous inkjet printing is still very much in use today. If you look at a yogurt or a can of tomatoes, they always have an expiration date, and often also a code that indicates the batch number. Manufacturers use a continuous inkjet printer to print that date. Continuous inkjet printers have the unique feature that they can print at a large distance. The droplets swiftly travel in space for a couple of centimeters and land on the metal or plastic surface, where they dry instantly. A program changes the label automatically every day or for the next batch. The process is so fast that it is very convenient for labeling one object after another in manufacturing lines. Given the importance of packaging in our society, it is not surprising that, in 2019, the continuous inkjet printing market accounted for more than one-third of the global inkjet printing market. Next time you check the expiration date of your milk or yogurt, try to picture the stream of flying droplets that wrote those numbers in less than a blink of your eyes.

But most printing systems based on deflecting continuous jets fell out of favor when "drop-on-demand" systems—which do not deflect the inkjet and only turn it on for printing—were invented in the 1970s by S. I. Zoltan[3] and by E. L. Kyser and S. B. Sears.[4] These systems, commercialized by Siemens and Silonics in the late 1970s and 1980s, ejected the droplets through a small—not microfabricated—aperture when the user applied a voltage pulse to a material that expanded electrically (a so-called piezoelectric ceramic). Although microfluidic principles were the basis of these early papers or patents, the word *microfluidic* does not appear in any of them.

Working at Canon in Japan, Ichiro Endo was the first to realize in 1977 that a bubble could form in ink by heating, thus displacing enough ink in a reservoir to eject a droplet. While initially working on piezoelectric ink displacement, he was suddenly inspired by a hot soldering iron that had accidentally touched a syringe full of ink, causing the ink to bubble up. In a few days, Endo built a prototype that later became the basis for Canon's BubbleJet printers. John Vaught at Hewlett-Packard in Corvallis, Oregon, independently invented a very similar process in 1978 based on heating the ink with thin-film resistors (which was dubbed ThinkJet at HP). When the two teams found out about each other's work, in an unusual move of industrial collaboration, the two companies signed a cooperation agreement.[5]

Manufacturing reliable devices proved much more difficult than building a few prototypes. To control the size and trajectory of the droplets, the engineers capped the ink reservoir with a small orifice. In the first prototypes, they created the orifice by punching or laser milling. However, it soon became evident that making precise, multiple holes required a more precise and reliable technique. In the beginning, HP engineers built metallic orifice plates using nickel electroforming.[6] This complex MEMS process uses photolithography and electrochemical deposition to form a nickel foil against a photolithographic mold. But the orifice plates had to be manually mounted, which added costs, and integration with electronics was a supplementary step.

In the meantime, two engineers at IBM, Ernest Bassous and Lawrence Kuhn, thought about a radically different way to improve drop-on-demand inkjet printers. In 1973, they read about Jim Angell's group at Stanford, building the first biomedical MEMS pressure sensors in silicon.[7] What caught their attention was the shape of Angell's sensors, which contained inverted pyramidal pits. Bassous and Kuhn envisioned that the pits could be used both as reservoirs and nozzles to eject ink droplets in an array configuration for parallel printing. As opposed to the HP orifice plates, which had to be electrochemically grown—a very lengthy process—Angell's group had used the topologically opposite approach to achieve their pits: chemical etching. To etch the pits, the Stanford researchers had used a hot solution of a commonly available, very caustic chemical named potassium hydroxide (KOH). It seemed relatively simple, and the material—silicon—was very familiar to IBM researchers building microelectronics next door, so for the same price Bassous and Kuhn could integrate electronics into their nozzle devices.

JAMES ANGELL AND THE STANFORD GAS ANALYZER

Every scientific field has an angel architect—a researcher that built the first scaffold and elevated the field to a height where everybody can see it. The angel architect of the MEMS field is James Angell (pronounced "angel" and known as Jim), professor of electrical engineering at Stanford. His work on MEMS and the new field of integrated circuits directly led to the first microfluidic chip, utilizing gas as its fluid.

In 1964, Angell heard an intriguing suggestion from a neurologist visiting Stanford's Integrated Circuit Laboratory (ICL).[8] Could this technology be used to make things other than electronics, like smaller microelectrodes for brain recordings? In 1965, an electrical engineering PhD student from Illinois by the name of Kensall Wise joined his lab. Most interestingly, he had worked for a year in the Digital Device Integration Department at Bell Labs in Murray Hill, New Jersey. Angell told him about the idea of the silicon probes. Wise remembered that people at Bell Labs had been working on techniques for silicon etching, so he spent the summer of 1966 back at Bell Labs to learn them. By 1970, Wise was routinely making neural probes in silicon with the diameter of a small (solid) needle—the first microelectromechanical system or MEMS.[9] Wise eventually became a professor at Michigan, where he built a stellar research program based on his neural probes. Wise's "Michigan probes" are now used all over the world.

MEMS were faster, smaller, and lighter, so they quickly caught the attention of NASA's engineers who wanted to reduce the weight of the onboard instruments on spacecraft and put life-monitoring, tetherless instruments on the astronauts. The records show that the NASA Ames Research Center (a few miles southeast of Palo Alto, California, where Stanford is) eagerly started awarding many research contracts to these MEMS pioneers around that time.[10] It was planetary biologist Glenn

Carle, likely along with other scientists in the upper management such as Harold "Hal" Sandler and Nigel Tombs from NASA Ames, who contacted Angell in 1971 with a specific request: Could a miniature gas chromatography column (box 2.1) be microfabricated in silicon?

The NASA engineers visiting Angell knew that gas chromatography equipment consisted of a bulky steel column ranging from half a meter to 3 meters long—not something astronauts could take with them in a rocket. The NASA visitors added that they had funding for it and that they were interested in a surface analysis instrument for the upcoming Mars lander for the Viking mission. This project was already the second that NASA had commissioned from Angell—the first had been a biomedical pressure sensor,[11] based on a thin piezoresistive diaphragm, that Angell had just completed. By chance, Jim Angell was contacted right then by Steve Terry, a master's student. "I had taken a couple of Angell's courses, and I had liked him as a teacher, so I wanted to know if he had funding for me to do PhD research in his lab," Terry vividly remembers. Terry, now in his seventies, seems to enjoy recollecting his student years in detail, his memory and good humor intact, as he sits in his house in Palo Alto not far from where he went to school. Steve Terry decided to join the lab and worked with then-postdoc Ken Wise and two others in the group.

BOX 2.1
Chromatography

Chromatography is a process for separating the components of a mixture. The name comes from joining the Greek words *chroma* (color) and *graphos* (recording): Italo-Russian botanist Mikhail Tsvet invented and named the technique in 1900 during his research on plant pigments. Chromatography can separate both gas and liquid mixtures. In traditional chromatography, a researcher fills the column with a stationary phase of, say, silica beads, and passes a mobile phase (a solvent plus the sample) through the column. The various components of the sample, depending on their molecular size and shape, go through the packed beads at different speeds, effectively separating them.

The separation performance of a chromatography column increases with its length, but nobody said the column had to be straight or could not be thinner. As part of his PhD research, Terry showed that he could coil a 1.5-meter-long chromatography column into a 5-centimeter-wide planar silicon wafer. To that end, he designed the "column" as a 200-micron-wide, 30-micron-deep spiraling microchannel etched in silicon (figure 2.2).[12] Researchers pack traditional chromatography columns with a porous matrix formed of, for example, silica beads—which then collide with the gas molecules—to slow down the various components of the gas mixture (see box 2.1). Terry simply relied on the gas molecules to collide with the walls of the microchannel, and in doing so, slow down according to their molecular weight. "I tried first with a 10-meter-long, 20-micron-deep channel, but I could not push the gas through the other side," recalls Terry with a smirk, as if the failure itself were a matter of amusement. He quickly explains that the failure pointed to how he could improve the device: the average free path of the molecules in the gas (the length that a molecule travels without colliding with another molecule) was longer than the smallest channel dimension, so there were not enough gas molecules to transmit the pressure. A shorter, wider, 1.5-meter-long channel worked well.

It seems a bit puzzling how a tiny, flat chip—now known as the *Stanford gas analyzer*—could outperform its giant predecessor, the chromatography column, both in terms of speed and overall device size. The key was that analyzing a microscopic amount was all that was needed. Terry's device separated a small sample—a few nanoliters—from a mixture of hydrocarbon gases in less than ten seconds. The benefits of scaling down were enormous: the performance (here, speed and device size) augmented with the square of the reduction factor, and other elements could be integrated at no extra cost.

Our body uses this microfluidic strategy of rapid, miniature sampling as well, such as in the hypothalamus, a tiny organ the size of an almond located in the brain. The hypothalamus is responsible for regulating key processes such as body temperature, hunger, thirst, blood pressure,

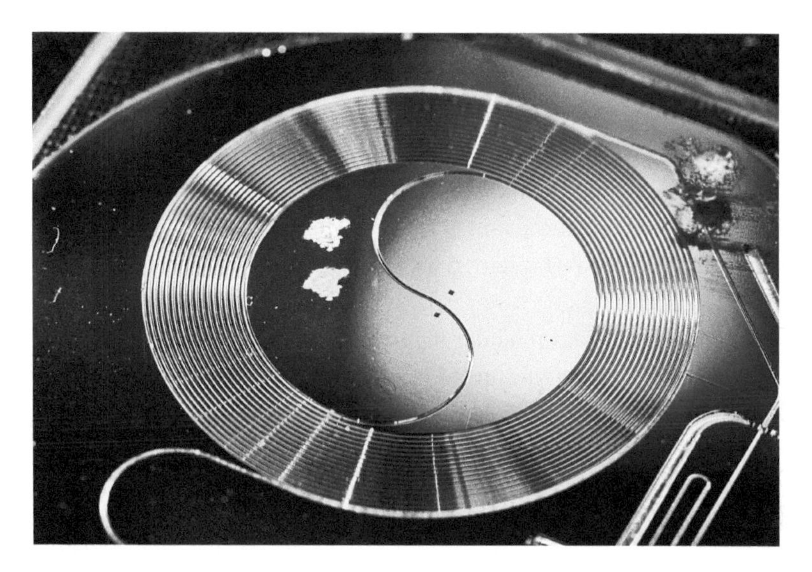

Figure 2.2
The planar chromatography "column" designed by Steve Terry for his gas analyzer. The spiral channel is 30 microns deep, 200 microns wide, and 1.5 meters long, occupying a 5-centimeter-wide circular area.
Image courtesy of CBMS (Okaar Photography).

fatigue, sleep, and sexual behavior, among other critical functions. A small network of blood capillaries perfuses parts of the hypothalamus. The hypothalamic neurons "read" relevant information from these capillaries, and they release necessary hormones into them to help maintain a balance of body functions. The hypothalamus is like a microfluidic chemical analyzer and Nintendo controller. By using this small-analyzer strategy, our body can react faster.

The design and integration of the output sensor was performed by PhD student Hal Jerman, who had joined the project in 1975. The sample (a mixture of gases) was pushed through the column by a neutral carrier gas (typically helium) and each component of the mixture came out through the column at a different time. The output gas stream from the column was passed over an electrical resistor that measured the thermal conductivity of the gas, which was related to the concentration of sample vapor in the helium carrier gas. Jerman had done a master's at Caltech with Carver Mead—a pioneer of modern microelectronics—on integrated circuit design, electronics, and solid-state physics. He initially had wanted to take a job at Intel when they were a single-building company in Santa Clara, but Mead convinced him to get a PhD instead. Jerman half-jokingly regrets that "a friend who started with Intel at the time did very well with his stock options."

The final design was modular (figure 2.3). The silicon device was simply a manifold carrying features like channels and the valve seat and to which external actuators could be attached (figure 2.4). This design was more practical because the user could replace, repair, or improve the parts, but it would also take much longer to assemble. Terry had to come up with

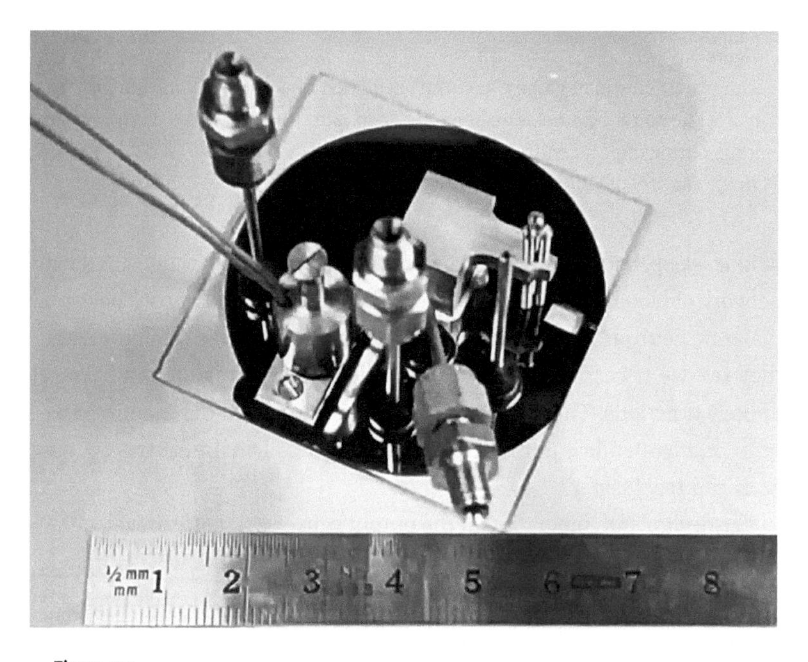

Figure 2.3
The Stanford gas analyzer.
Image courtesy of Steve Terry.

Figure 2.4
Detail of the fluidic circuitry underlying the Stanford gas analyzer.
Image courtesy of CBMS (Okaar Photography).

the microvalve seat actuated by an external solenoid, so it became possible to inject ~1 nanoliter volumes of gas—the first microfluidic valve ever reported. Steve Terry still remembers that one of the biggest challenges was to achieve a hermetic bonding—leaks in gas chromatography are critical—and that "someone, at a party, suggested that I try anodic bonding with Pyrex glass—it worked!"

Unfortunately, in the end, the Stanford gas analyzer was not the instrument chosen to fly on the Viking mission, and challenges in the assembly and packaging made its commercialization difficult. But the legacy endures. In April 1983, Jim Angell and Steve Terry, along with Philip Barth, coauthored a cover story in *Scientific American* entitled "Silicon Micromechanical Devices" that opened the eyes of the public to the MEMS field.[13] Kurt Petersen's influential review article "Silicon as a Mechanical Material,"[14] published in 1982 (cited almost 4,300 times as of 2020), prominently reviewed Angell's work and, perhaps as importantly, reproduced a figure from Terry's paper illustrating the different geometries that result from silicon etching (figure 2.5). Petersen's review has inspired tens of thousands of MEMS engineers worldwide; they have studied Terry's figure and built other designs based on its principles. Perhaps Hal Jerman's friend got better stock options, but Terry, Jerman, and Angell have left an imprint in the collective memory of the field.

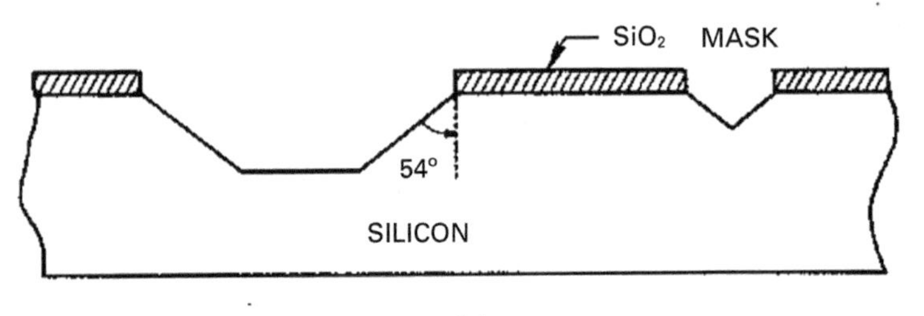

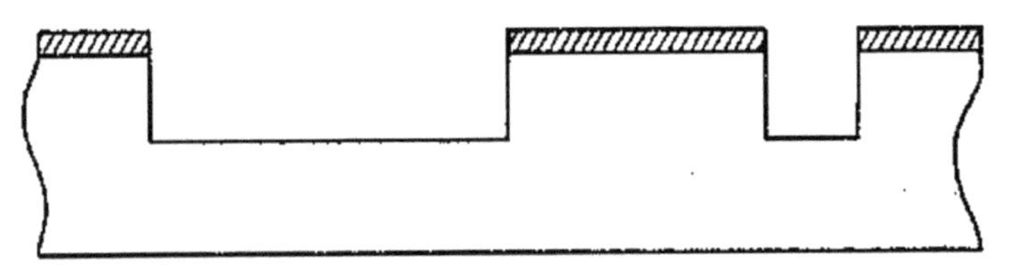

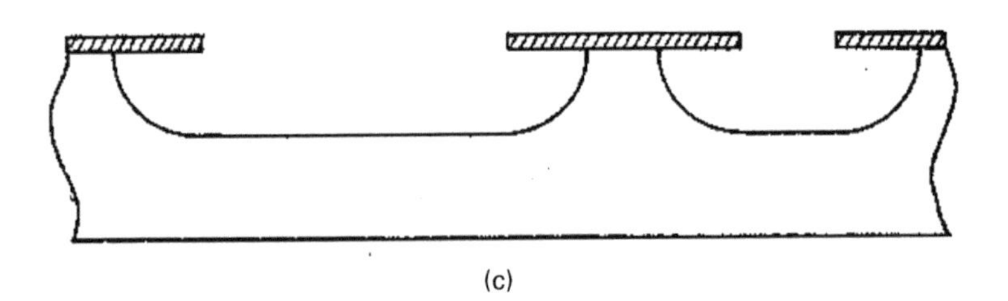

Figure 2.5
Cross sections of grooves etched in silicon with (a) KOH etchant in silicon {100}, (b) KOH in silicon {110}, and (c) HF-HNo3 etchant.
Image courtesy of Steve Terry.

THE INKJET NOZZLE ARRAY

The Stanford gas analyzer was the first MEMS-based microfluidic device in history and was the inspiration for subsequent MEMS-based liquid-handling devices. In 1977, Angell developed a MEMS capacitive pressure sensor[15] using the new KOH etching techniques that would catch Kuhn's and Bassous's eyes.

From Angell's work, Kuhn and Bassous cleverly designed inkjet nozzle arrays as a set of KOH-etched pits in 1977.[16] The chemical etching of silicon had been studied extensively for more than a decade. If one dipped a silicon wafer in a bubbling-hot KOH solution, the silicon would be gone in less than an hour, etched away by a chemical reaction between KOH and silicon. The rate of dissolution depended strongly on the crystal orientation of the silicon. Crystallographers designate the crystal planes or orientations as {100}, {110}, and {111}. A wafer can be grown and polished in the {100} direction (a "Si {100} wafer"), but inside its crystal, it still contains other crystallographic planes. The {110} orientation etched one hundred times slower than the {111} orientation.[17] The higher packing density of atoms in the {111} planes made them more resistant to etching than the {100} and {110} planes. That meant that a Si {100} wafer covered in silicon oxide (which is resistant to KOH etch) and patterned to expose square windows of the top silicon surface would form pits with four self-terminated mirror-finish {111} faces, each angled at exactly 54.7°, resulting in an inverted pyramidal pit [figure 2.5(a) and figure 2.6]. Using wafers that were only under 200 microns thick, they were able to obtain square openings that were just over 30 microns on the bottom face of the wafer.

The regularity of the pits caught Kuhn's and Bassous's attention. Thanks to the crystalline nature of silicon, the pyramidal walls were always at the same angle. The mechanism for the KOH selectivity was not known then, but Kuhn and Bassous saw an excellent utility for the pits. They used each pit as the interior of a nozzle and, by fabricating a grid of pits, created large nozzle arrays that a user could operate in parallel. With this first device, they could shoot eighty-four thousand drops per second, each

Figure 2.6
KOH-etched pits in a silicon wafer (by Andreas Manz).
Image courtesy of CBMS (Okaar Photography).

droplet flying at a speed higher than a meter per second.[18] To deliver ink on demand, next to the pitted reservoirs, they fabricated an electrode that they used to heat the ink to its boiling point; the heat created a tiny bubble that expanded the fluid and ejected a droplet through the nozzle. They could turn the heat on and off thousands of times per second because silicon would quickly absorb the remnant heat.

Canon commercialized this type of inkjet process under the name of BubbleJet. Another method for ejecting the fluid used a flat piece of piezoelectric material that expanded when electrically excited. Thus, a sudden, brief burst in voltage applied to the piezoelectric material caused the ejection of a tiny amount of fluid through the nozzle (figure 2.7).

Notably, the top surface next to the piezoelectric slab or the bubble-generating electrode could be patterned with the necessary microelectronics to control the delivery of the ink. It scaled remarkably well, as most microelectronics and MEMS devices do: it was as easy and cheap to fabricate one nozzle as it was to fabricate hundreds on a wafer with the

Figure 2.7

Sequence of high-speed micrographs (from left to right) depicting the ejection of two droplets from the same nozzle. The diameter of the nozzle is 30 micrometers, and the time between each image is three microseconds. For this type of printhead, droplets can typically be produced with a volume of 1–32 picoliters, the jetting frequency ranges between 10 and 100 kilohertz, and the final droplet velocity ranges from 5 to 10 meters per second.

Image courtesy of Arjan Fraters et al. See A. Fraters, M. van den Berg, Y. de Loore, H. Reinten, H. Wijshoff, D. Lohse, M. Versluis, and T. Segers, Inkjet Nozzle Failure by Heterogeneous Nucleation: Bubble Entrainment, Cavitation, and Diffusive Growth, Phys. Rev. Applied *12, 064019 (2019).*

necessary electronics. Canon and Hewlett-Packard rapidly adopted the technology.[19] The familiar inkjet printers spit drops of ink on demand as the nozzles scan over the paper surface—with such precision that they can print text and images in different inks and dots of various sizes, giving you the impression of smooth color gradients. It all happens much too quickly for you to see because the droplets are a few thousandths of a millimeter in diameter and travel at 10 meters per second. That means that a droplet flies through the 1-millimeter air gap between the nozzle and the paper in a fraction of a thousandth of a second. Some inkjets spit out liquid ink, but some can print wax, giving a shiny photo appearance. Some 3D printers are inkjet printers that use photopolymers to build a 3D object droplet by droplet (chapter 9).

Significantly, microtechnology benefits from cost reduction by batch fabrication and integration. The inkjet nozzle is a perfect example: the smaller the nozzle, the better it performs, and more nozzles can be packed in one wafer, so the production costs per unit become cheaper. The nozzle array was intended originally as part of a permanent printhead, but most inkjet printers now incorporate a disposable ink cartridge. This way, manufacturers no longer need to deal with frustrated customers calling about clogged printheads. In 2019, the global inkjet printer market was valued at more than $34 billion with a 5 percent growth rate. More than forty years after Bassous and Kuhn invented the silicon inkjet nozzle, this microfluidic component has become an essential part of millions of printers and 3D printers at our homes and offices.

FROM MOLECULAR GASTRONOMY TO DROPLET MICROFLUIDICS—AND BACK

As important as printers are for our lives, ink droplets constitute an insignificant fraction of the droplets you typically use every day. You ingested millions of droplets today if you had milk or salad dressing, but you do not need to call your doctor: that is the normal composition of a salad dressing or a glass of milk. These microfluidic mixtures of oil and water in the form of tiny droplets are called *emulsions*. Water and oil do not mix (figure 2.8), but you can vigorously whisk a water-based solution like vinegar together with oil and form salad dressing—a type of emulsion made of a suspension of vinegar droplets into oil. This emulsion is not stable because, as the vinegar droplets collide with each other, they prefer to coalesce rather than to stay apart. Slowly and surely, the oil and vinegar will separate no matter how much whisking you have put into it—even if you use a blender.

The process by which these small droplets form is a microfluidic equilibrium of interfacial forces. Your whisk simply breaks the water–oil interface and allows the emulsifier (box 2.2) to intercalate itself between oil and water (or vinegar or lemon). The more you whisk, the smaller the droplets, the more homogeneous the sauce, and the more stable the droplets. Chocolate is another example of an emulsion, in this case of milk and cocoa butter.[20] Milk itself is a complex emulsion of water, small submicron

Figure 2.8
Stirring of food dyes into olive oil generates an emulsion, a suspension of droplets in oil.
Image courtesy of Andrew deMello, ETH Zurich.

particles of a type of proteins called caseins, and butterfat globules. You can break the emulsion by adding lemon juice (curdling), which causes the casein particles to separate from the liquid and coagulate. If you have enjoyed any or all of these foods, you have been enjoying the taste of microfluidic droplets.

Droplet microfluidics have enabled the miniaturization of chemistry glassware to one quadrillionth its size. The large size of the reaction vessels

BOX 2.2
Emulsifiers

To stabilize an emulsion, you need an *emulsifier*, a type of surfactant. Surfactants are molecules that insert themselves at the surface of a fluid, reducing the surface tension. Emulsifiers coat droplets within an emulsion and prevent them from joining together, or "coalescing" (figure 2.9). Mayonnaise is also a mixture of oil and vinegar, plus egg yolk. If you whisk or blend the three components, tiny droplets of oil are visible but do not coalesce. Egg yolk has lecithin, a fatty substance that is soluble in both the oil and the vinegar, basically covering the droplets and preventing them from coalescing. Hollandaise sauce, like mayonnaise, also uses the emulsifying properties of egg yolk, but instead of oil and vinegar, you use butter and lemon, respectively.[21] Thanks to the lecithin in egg yolk, if you look under the microscope at mayonnaise or hollandaise sauce—or any other sauce that contains egg yolk, such as Béarnaise sauce—you will see stable droplets. There are other natural emulsifiers. The emulsifying properties of garlic led to the sauce called aioli[22] (from Catalan *all i oli*, or garlic and olive oil; the French recipe adds egg yolk to shorten the emulsification process—a shortcut that is scorned at by Catalan purists).

Figure 2.9
This image seems like a science-fiction illustration of a turbulent galaxy, but it is a real demonstration of how surfactant accumulates at droplet boundaries—where surfactant is supposed to be. The surfactant has been labeled with a green fluorescent dye so it can be observed. The beads are moving and leave a cosmic-looking trail of fluorescence in the camera. *Image courtesy of Andrew deMello, ETH Zurich.*

(beakers, flasks, test tubes) and fluid transfer glassware (pipettes, funnels, burettes) traditionally used by chemists had been a challenge to performing many reactions in parallel, to reducing the cost of the consumed reagents, and to speeding up the reactions. Chemists have been interested in emulsions for a long time, but the technology for making homogeneous droplets in large quantities dates from just twenty years ago.

A chemist looks at a droplet and sees a microscopic test tube, so an emulsion looks to them like a collection of billions of infinitesimal test tubes. Since emulsions can be created merely by whisking and other stirring methods, several chemistry groups had started studying chemical reactions confined to droplets.[23] However, the droplets were not homogeneous in size. Here is where microfluidic engineer Mitsutoshi Nakajima at the National Food Research Institute in Tsukuba ("Science City"), one hour north of Tokyo, in Japan, thought of injecting oil into water (and vice versa) through small apertures inside a silicon microchamber. In 1997, he observed that, due to the instability that occurs at the oil–water interface, the device could create uniform droplets at high throughput.[24] Four years later, Stephen Quake's group realized that droplets could be formed in a very simple, PDMS-molded device that brought together an oil stream and a water stream in a T-junction (figure 2.10).[25] Shelley Anna

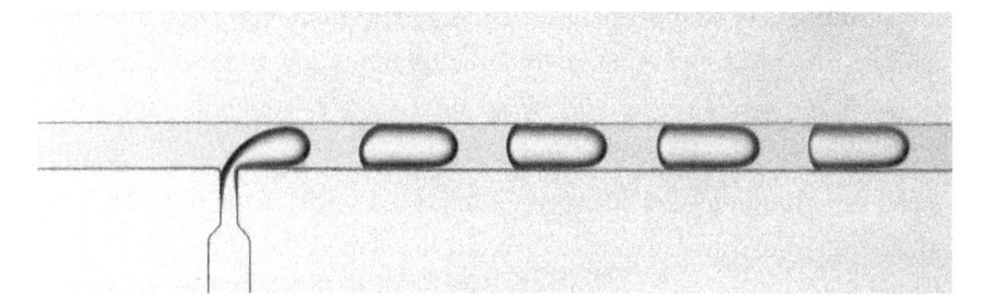

Figure 2.10
Generation of droplets in a 50-micron-wide microchannel using a T-junction. A water stream (colored blue) is continuously injected at a 90-degree angle into a mineral oil stream (colored yellow). Flow is from left to right. The water stream breaks up spontaneously into droplets that are transported downstream by the oil flow.
Image courtesy of Andrew deMello, ETH Zurich.

and Howard Stone presented a conceptually equivalent "flow focusing" PDMS design in 2003.[26] Both PDMS devices are among the ten most-cited papers in the history of microfluidics.

Almost two decades later, *droplet microfluidics* has grown into a vibrant field of research.[27] These devices (usually made of PDMS) generate and handle picoliter-volume droplets at kilohertz rates. That means that, in one second, a microfluidic device can make and process up to ten thousand droplets—but they are so small that it can take almost a *trillion* droplets to fill a 1-liter bottle. The droplets can be merged (figure 2.11), fused, mixed, reacted, split, sorted, and analyzed at these rates, allowing millions of different combinations of chemical reactions to be run in less time than it takes you to finish your coffee. Researchers have used microfluidic droplets to push the boundaries of chemistry.[28] For example, droplets have allowed for optimizing protein crystallization; researchers produced 1,300 crystallization trials in twenty minutes using 10 microliters—a drop smaller than a tear—of protein solution.[29] Approximately one hundred million enzymes (a special type of proteins that help accelerate biochemical reactions) were screened in ten hours using 150 microliters of reagent—an experiment used to direct the "evolution" of enzymes.[30]

Droplets have also revolutionized the way scientists amplify and analyze the genetic material (such as the DNA that makes up our genes) of large numbers of single cells. By decoding (or sequencing) DNA, researchers can investigate whether there are alterations (or mutations) responsible for a given disease, and by sequencing a related molecule called RNA, they can reveal which genes are being switched on and provide a full description of the "identity" of the cell. In 2015, a team led by Steve McCarroll and David Weitz at Harvard developed the *Drop-Seq* method (short for "droplet sequencing"), based on co-encapsulating a single cell and a unique bead into each droplet.[31] The bead was "barcoded" with unique DNA molecules serving as tags and allowing for tracking the identity of the cell during RNA sequencing. To demonstrate the technique, the team processed approximately forty-five thousand cells from the retina through the microfluidic device, identified all ten expected retinal cells

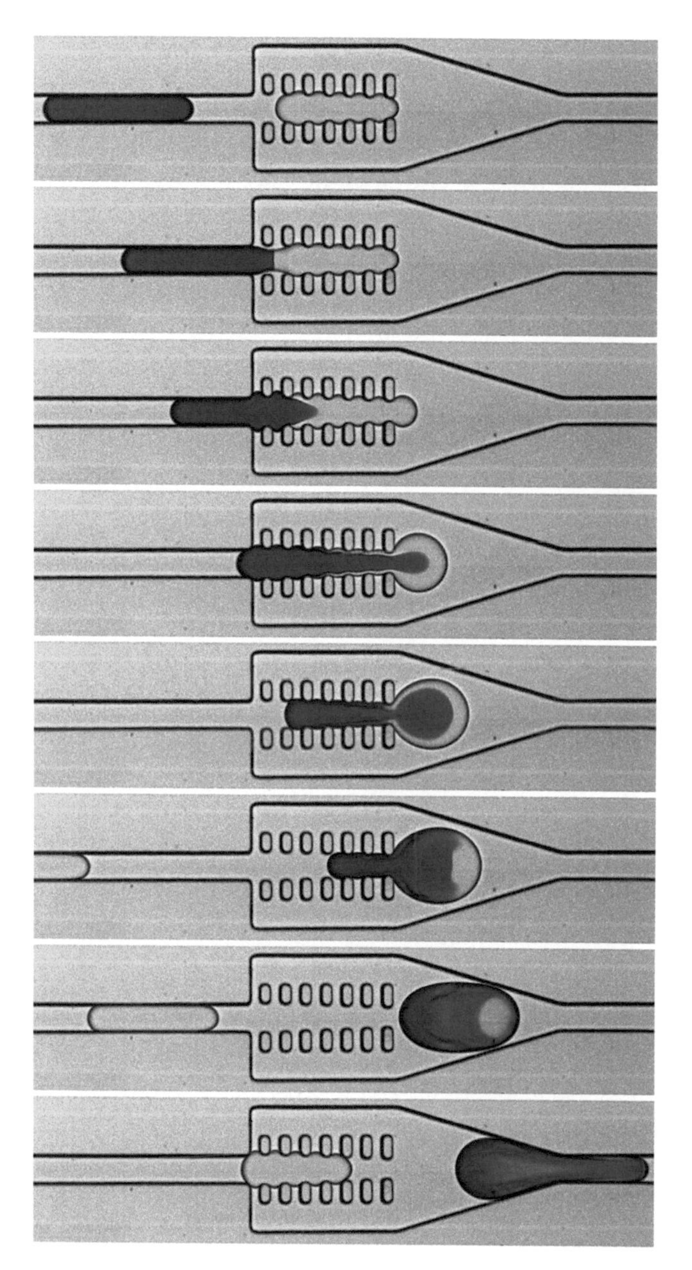

Figure 2.11

High-speed filming of the merging of two droplets. The clear droplet gets trapped upon entering the chamber because the oil goes around the droplet until the second (dark) droplet collides against it from behind. The collision merges the two droplets and provides enough energy to leave the trap. The whole sequence takes just over one hundredth of a second to complete. *Image courtesy of Andrew deMello, ETH Zurich.*

for which molecular markers exist, and—as a happy bonus—was able to distinguish as many as thirty-nine subtypes of cells. Importantly, the team wanted to make sure that any lab could build its own system so they made Drop-Seq freely available as open source. As a result (and because Drop-Seq has brought the cost of sequencing down to about six cents per cell), hundreds of researchers have now implemented it in their labs. In 2017, Adam Abate's group at the University of California, San Francisco isolated, fragmented, and barcoded the genomes of more than fifty thousand bacteria with droplets in just one experiment for parallel sequencing.[32] Similar experiments with tumor cells in droplets enable the identification of cancer-related mutations in thousands of individual cells. The numbers are mind-numbing.

Droplets caught the attention of Andrew deMello for their potential in chemical synthesis (box 2.3). "I was interested in using microfluidics for synthesis, to make things," he says. In a 2002 paper published with his brother John, they used a continuous-flow microfluidic reactor to synthesize nanometer-scale crystals in a rapid and controllable manner. Although this was a nice advance, it taught deMello a hard lesson: because the flow speed in microchannels follows a parabolic profile—it is zero at the walls and fastest at the center—crystal growth could not proceed at a constant rate. Additionally, the large surface-to-volume ratios of microchannels led to nanocrystal "seeding" on the walls and ultimately channel blockage.[33]

Droplets seemed like the perfect solution to these problems. "For us, initially, droplets were all about forgetting the surface, and all of the associated problems," recalls deMello. Droplets allowed researchers to perform discrete reactions in a highly controllable way. "You suddenly go from an environment where you worry about fouling and blockage and chip failure to one where you can form your droplets. If you have a big enough reservoir, you just do this for days, and days, and days," says deMello (figure 2.13). "Fouling" is a common problem in the microsensor field whereby the solid surface of a sensor becomes coated (or "fouled") with proteins that block the active sensing surface. A droplet is immune

BOX 2.3: BIOGRAPHY
Andrew deMello

DeMello grew up in Harrow, a green and leafy suburb in the northwest of London. His father, Joe deMello, was an IBM machinist who had emigrated from Goa in 1961 when India took the former Portuguese colony back in a brief armed engagement that lasted thirty-six hours. Joe ended up in London in the mid-1960s, where he met and married Norma. They had two children—Andrew (born in 1970) and John ("definitely smarter than me," says Andrew), both of whom would become professors at Imperial College London, despite the family's lack of academic background. Andrew credits his high-school teacher David Weedon for his passion for chemistry—"and my math skills were never like my brother's," he adds. Andrew obtained a BS in chemistry from Imperial College; John studied physics.

It's not like Andrew pursued the easy route. He did a PhD in molecular photophysics—a subfield of chemistry that studies how light interacts with molecules—at Imperial College. At the end of his PhD, in 1995, he applied for a postdoc on laser studies in Richard Mathies's lab at the University of California, Berkeley. Mathies responded that he did not have funding for laser studies, but he hired deMello for "a project that uses microchips to look at DNA."

"I was in the right place at the right time because Rich had just got this wonderful graduate student named Adam Woolley, who had done the first single base-pair resolution DNA separations on a chip, and I got to learn from him" (chapter 3). After his postdoc, at just 25 years old, deMello took an assistant professor position at the University of East Anglia in London. However, Imperial College's Department of Chemistry quickly snatched him up a year later along with microfluidic pioneer Andreas Manz (chapter 3). In 2011, deMello moved to ETH Zurich, one of the cradles of microfluidics. Droplet microfluidics is also a field that produces movies and images of arresting beauty (figure 2.12). "We are lucky that we work in a field where there is a lot of visual capital to be made," says deMello, one of many scientists who believe that the words and pictures used to communicate the results are almost as important as what you do in the lab.

(continued)

BOX 2.3 (continued)

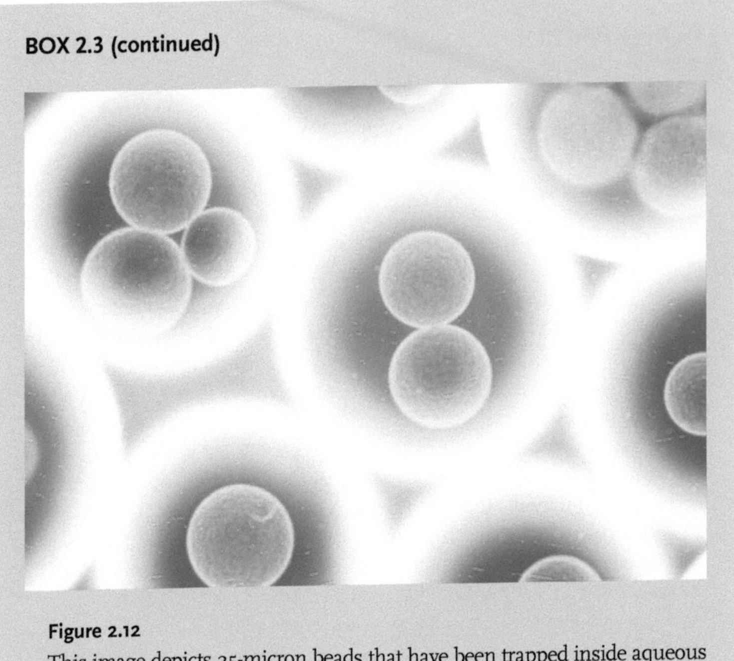

Figure 2.12
This image depicts 25-micron beads that have been trapped inside aqueous droplets by a process of merging different droplets separated by oil. Reactions that grow crystals, hybridize DNA, or stimulate live cells are also possible inside single droplets.
Image courtesy of Andrew deMello, ETH Zurich.

to fouling because its surface is liquid. "Synthetic chemists have for over two hundred years used test tubes and flasks for chemistry only because of ergonomics. A test tube fits in your hands, you can shake it, you can look into it. But there is no reason why you do that apart for convenience," he explains. That is why, in part, deMello advocates for simplicity of design in microfluidics—complex designs tend to fail and not work better. And droplets can be generated with very simple, easy-to-understand designs. For example, engineers can use a T-junction to bring oil and water together to form oil-in-water droplets with bespoke properties.

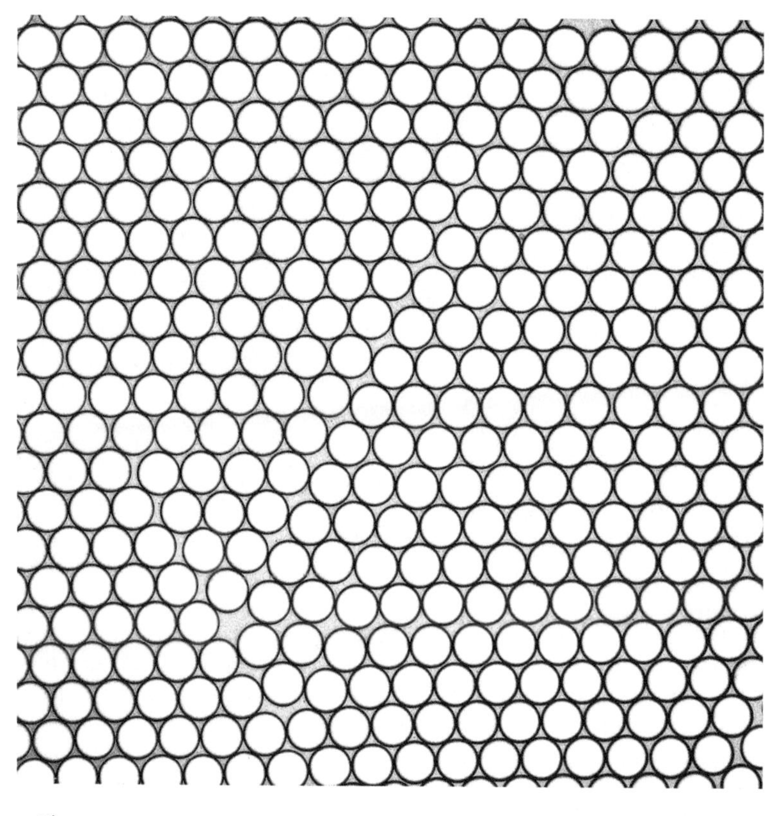

Figure 2.13
Microfluidically generated water droplets (white) surrounded by oil (colored). The diameter of the droplets is 70 microns.
Image courtesy of Andrew deMello, ETH Zurich.

"It's really obvious when you think about simple scaling laws that things should work better, more quickly, and in a more controlled manner when you shrink them down," he adds. The deMello lab has produced many elegant designs, such as a device that automatically generates a set of dilutions in sequential droplets to run a DNA-binding assay.[34] Over the years and using his laser expertise, deMello's lab has devoted much effort to the optical readout of droplets (figure 2.14). "How can we look into

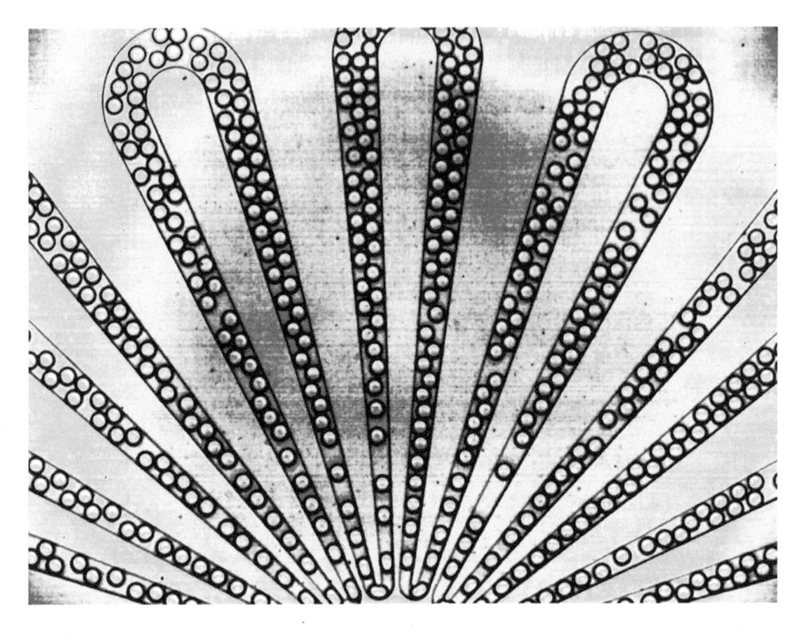

Figure 2.14
Droplet microfluidic device for studying the unfolding kinetics of an enzyme. The flowerlike arrangement of the microchannels seeks to concentrate droplets toward the center of an optical detector that uses ultraviolet light to measure enzyme unfolding. The droplets are about 90 microns in diameter and travel the length of one "petal" of the flower from the outside to the inside in about one second.
Image courtesy of Alessia Vallois and Andrew deMello, ETH Zurich.

these small volumes and pull out as much information as we can?" he wonders. "That was always a really big challenge, because however efficient your microfluidic system is, if you cannot see what you've done, then you are losing all this valuable information," he reckons.

* * *

One of the most remarkable stories of this field is the interaction between Dave Weitz, Harvard University physics professor and droplet microfluidics expert, and Ferran Adrià, a renowned chef from Catalonia,

Spain. Adrià is not just a great chef—most top chefs credit him as being the best chef *ever* for the brilliant creativity of his dishes over the years.

In November 2011, as I was strolling around Cambridge after giving an invited talk at a microfluidics conference, I was startled to notice Adrià walking directly toward me as recognition dawned on both of our faces. "Ferran, what are you doing here?" I asked, amazed at the odds of meeting him again there (see box 2.4), and excited like a teen in front of a rock star. Pedestrians around us seemed oblivious to Adrià's worldwide fame. In our shared hometown of Barcelona, by now, he would be surrounded by a mob. "Dave Weitz has invited me to give a course at Harvard," he replied with his signature facial expression—a mischievous smile and eyebrows raised—to share his incredulity. I then remembered that Weitz had invited Adrià to Harvard in December 2008 to give a talk about science and cooking. The talk, attended by several hundred people, had been a big hit, with large lines forming out the door.[35] As examples of his deconstructivist cuisine, Adrià demonstrated "caviar" of melon and "pasta" made of ham.[36] Because of the success of that first talk, in 2010, Weitz and other Harvard instructors started the popular class "Science and Cooking,"[37] which is still taught. "It's a way to convince people that it's fun and that there is a lot of stuff we understand from a scientific point of view that chefs exploit," Dave Weitz, who co-teaches the course, has said.[38]

"The attraction I had originally to Ferran is that he thinks like a scientist," points out Dave Weitz—a remarkable compliment given that Adrià has no science education beyond high school. Weitz has been interested in applying droplets to analyzing single cells for a long time. In 2008, using traditional droplet technology, a large team led by Dave Weitz, Andrew Griffith, and Christoph Merten showed that it was possible to keep cells—and even worms—inside droplets for several days,[39] but the oil and the surfactant remained a concern for long-term cultures. "Ferran's spherification process is a perfect application of our multiple drop technology," adds Weitz. *Spherification* is a process that uses calcium ions to trigger the gelation of alginate, an edible hydrogel derived from seaweed (figure 2.16). The process was first patented in Britain in 1942

BOX 2.4: BIOGRAPHY
Ferran Adrià

Adrià not only *is* a genius but he also *looks* like the caricature of a genius—with frizzy gray hair crowning a broad forehead and thick eyebrows intensifying his already fiery eyes. Adrià was born in 1962 in the suburbs of Barcelona. When he was 18, he decided to quit his studies and started working in various restaurants to pursue his goal to become a chef one day. At twenty-two, he was hired by *El Bulli*, a small restaurant at the end of a winding road on a remote beach two hours north of Barcelona. Eighteen months later, he was promoted to head chef. The restaurant meteorically rose to fame under the explosive ingeniousness of Adrià. *El Bulli* got three Michelin stars, and *Restaurant* magazine awarded *El Bulli* the best restaurant in the world a record five times between 2002 and 2009. His "deconstructivist" culinary creations "provide unexpected contrasts of flavor, temperature, and texture. Nothing is what it seems. The idea is to provoke, surprise and delight the diner," he has said.[40] *Time* magazine chose Adrià as one of the one hundred most influential people alive. He has been on the covers of the *New York Times* Sunday Magazine, *Le Monde*, *Esquire*, and *Wired*. By the time it closed in 2011—because it was operating at a loss and Adrià did not want to raise prices for fear his restaurant would lose its charm—*El Bulli* was getting two million reservation requests—which entailed writing a detailed essay—every year, of which only eight thousand fortunate ones were accepted. I like to think that a mention of microfluidics in my essay helped my wife and I get a table in July 2010. At *El Bulli*, we met Adrià (figure 2.15), as he always went around the tables to meet his guests. He spoke with such passion for his culinary inventions, with so much artistic fire in his eyes—I thought this is what Leonardo da Vinci must have been like.

Figure 2.15
Ferran Adrià, between Lisa Horowitz and the author at the restaurant *El Bulli*, July 2010.
Image by Albert Folch, University of Washington.

by William Peschardt, a food scientist working for the firm Unilever, and demonstrated by the representatives of a Spanish food company to Adrià, who then brought it to fame and put it to creative uses at *El Bulli*, starting around 2003.[41] A search of the term *spherification* on YouTube retrieves thousands of videos from wannabe chefs that make little liquid balls of everything edible to entertain their guests. The process entails thoroughly mixing your liquid food of choice—say, a fruit punch—with alginate powder and then dropping it into the calcium-containing solution. After a few minutes, the outer layer of the alginate mixture has gelled, creating a soft liquid "bubble" that you can safely manipulate but also easily pop in your mouth for your enjoyment.

The spherification process is essentially a gentle packaging process for food. As such, it has found uses in droplet microfluidics and the biomedical field. Researchers have known since the 1980s that pancreatic

Figure 2.16
Spherified "olives" and "peanuts" at *El Bulli*. These edible constructs are gelatinous and can be used for food packaging, a property that has been applied to the encapsulation of live cells in microfluidic droplets.
Image by Albert Folch, University of Washington.

islets can be "microencapsulated" inside droplets of alginate. Pancreatic islets are small balls about a tenth of a millimeter in diameter of pancreatic cells that are responsible for insulin secretion in our body. If islets from one animal are implanted into a different animal, they keep secreting insulin for months without causing an immune reaction.[42] Doctors have also treated brain tumors by delivering genetically engineered cells encapsulated in alginate droplets.[43] However, in these early contributions, the cells were encapsulated using a dispenser, which limited the number of droplets that could be generated. The team led by Mitsutoshi Nakajima in 2005 produced another ground-breaking demonstration of droplet technology: that the alginate gelation process could be miniaturized and automated with a microfluidic droplet generator.[44] Others (including Weitz) have followed with various devices and applications for packaging molecules and cells into droplets.[45]

The idea is deceivingly simple. The researchers first flow the cells in an alginate solution in a microchannel. As the alginate stream joins an oil stream, it forms liquid alginate droplets in oil at a rate of hundreds or thousands of droplets per second. These droplets are then fused downstream with calcium-containing droplets that cause the alginate to gel. At the end of the process, the researchers can remove the oil—an undesired element since it is not compatible with cells in the long term—because it is no longer necessary to keep the droplets from coalescing. The process is more straightforward than traditional water-in-oil droplets because emulsifiers are no longer needed here. Interestingly, spherification was appealing to the palates of *El Bulli*'s visitors because it packaged foods into fun bubbles that exploded inside their mouths—but for microfluidic engineers, the opposite is true: spherification is a tool for the oil-free, safe packaging of molecules, cells, and tissues.

* * *

That something as small as droplets could impact our lives to such an extent is breathtaking. The inspiration came from nature, but the ingeniousness required to develop a device that can automate the printing of letters, pictures, and 3D structures by shooting ink droplets deserves applause. I find it poetic that engineers have found inspiration not only in mist and rain but also in the natural emulsions of foods to develop these ultrafast droplet-generating devices that have multiplied the yield of chemistry and genetics experiments exponentially. Could the next frontier for droplets be in drug manufacturing? Indeed, Patrick Doyle's lab at MIT has recently fashioned drug-loaded droplets with alginate capsules to tailor drug release.[46] Maybe in a few years, drugs will come microfluidically encapsulated as emulsions, the droplets popping to release their drug load inside of your cells just like *El Bulli*'s spherified foods pop inside the mouths of mesmerized customers.

3 DECODING THE HUMAN GENOME WITH MICROFLUIDICS

These days everyone and their mother seem to have their DNA—and even their dog's DNA—tested to find out about some disease or their ancestry. This DNA testing craze exists thanks to microfluidic devices developed about a decade before the advent of droplet-based sequencing that can rapidly sequence DNA.

Our DNA is a molecular blueprint containing the instructions with which our bodies are built in the womb and then repaired and operated daily. DNA sequencing technology has allowed scientists to decode the meaning of those instructions. This achievement has revolutionized our lives and occurred in the lifetime of the average reader of this book. The molecular structure of DNA was unveiled in 1953 by Francis Crick and James Watson, who obtained a Nobel Prize in 1962 for the discovery. Because academic researchers obtained the first DNA sequences using laborious methods in the early 1970s, they only decoded short sequences of our DNA at a time. More recently, rapid-speed *microfluidic DNA sequencers* have been instrumental in the sequencing of all the DNA ("genome") of numerous types and species of life, including the complete genomes of many animal, plant, and microbial species. This knowledge and capability has revolutionized many other fields and applications, including drug discovery, diagnostics, evolutionary biology, and virology.

Because each human has unique DNA, DNA sequencing has provided an essential tool that has dramatically improved the reliability of paternity tests and forensic science in the last thirty years. Incredibly, DNA can be isolated and amplified from a fingerprint, a hair, a speck of saliva, or a spot on a handkerchief. It is now easier to identify the victims of a catastrophe or a crime and rule out crime suspects. The Innocence Project, which started in 1992, has used DNA sequencing technology to exonerate 375 people who each wrongfully served an average of fourteen years in prison in the United States, twenty-one of whom were death row inmates; thus, a total of 5,250 human years have been spared the suffering of unjust imprisonment. Some companies such as 23andMe now offer direct-to-consumer personal genome testing, which has prompted many individuals to seek paternity testing, genealogical testing, testing of medically relevant genes, and even testing their pets' DNA to identify the breed and other inherited traits, such as health risks. This flurry of DNA sequencing tests would not have been possible without the advent of microfluidic sequencers in the 1990s under the auspices of the Human Genome Project.

THE HUMAN GENOME PROJECT

Genes—encoded in DNA base-pair sequences—are instructions for cells to function and reproduce themselves. In sum, they are responsible for how we inherit traits and disease-related mutations from our parents and ancestors. The ensemble of all our 25,000-odd genes—a total of 3.3 billion base pairs of DNA—is the human genome. To decode the whole sequence, in 1984 the US government started planning the Human Genome Project, which was finally launched in 1990. The initial goal was to find the genetic roots of disease and then develop treatments, such as identifying mutations that increase the risk for common diseases like cancer and diabetes. In March 2000, President Clinton announced that the genome sequence could not be patented and should be made freely available to all researchers.

Microfluidic chips played a fundamental role in decoding the human genome by providing quick and accurate DNA-based diagnostic tools, making the genetic information more useful than before. The first complete human genome sequence took more than a decade, and would have taken much, much longer without the aid of microfluidics. Now, researchers can analyze changes in the activation of genes during the evolution of a disease—such as during the growth of a tumor after a chemotherapy dose—that may lead to information-driven approaches to therapy. Medical technicians can sequence genes (or the full genomes) from placental cells and compare known variations with the parents to see if there is a risk of genetic diseases in the fetus. The Human Genome Project efforts—which were declared complete in 2003—extended across twenty universities and research centers in at least seven countries—the United States, Canada, the United Kingdom, Japan, France, Germany, and China.

At one of these universities, the University of California, Berkeley (UC Berkeley), chemistry professor Richard Mathies and his team developed the sophisticated microfluidic glass chips and improved fluorescence detection methods that made this high-throughput DNA sequencing possible. To understand how Richard Mathies was able to develop glass chips for the Human Genome Project, we need to rewind to 1985, when the Swiss chemist Andreas Manz was finishing his PhD at ETH Zurich.

ANDREAS MANZ, THE ARCHITECT OF MICROFLUIDICS

Inadvertently or by instinct, Andreas Manz became a human connection between the budding MEMS communities of Japan and Switzerland. Switzerland, with its watch industry pushing for MEMS developments, and Japan, with its microelectronics industry interested in integrated MEMS sensors, had become two brewing MEMS cauldrons waiting for microfluidics to cook up. Manz grew up in Rüti, a small Swiss town, half an hour east of Zurich. As an adolescent, he developed a fondness for art and moths, but seeing his parents struggle as they worked hard in modest jobs, he knew he did not want that for himself. For college, Manz considered architecture and biology, but chemistry won. He graduated in 1982 with an analytical and organic chemistry degree at ETH Zurich, at the tip of Lake Zurich. Manz also wanted independence, so for his PhD (also at ETH Zurich), faced with three choices, he picked an advisor who was not around much: Wilhelm Simon, who worked on ion-selective electrodes and microelectrodes. One day, in 1985, upon returning from a trip from China, Simon walked into the 4 p.m. coffee break at ETH and asked: "Who wants to go to China?" Young Manz immediately said, "I'll go." But the same day, Simon told him about a more attractive opportunity at Hitachi in Japan. Simon could arrange a postdoctoral project postdoc in Hiroyuki Miyagi's lab in the field of clinical diagnostics and analytical chemistry. Did the country matter? Manz's eyes were set on the future. In January 1986, Manz arrived in Hitachi city, a coastal town just an hour and a half north of Tokyo.

Even though he was a new employee, Hitachi let him attend the Transducers '87 conference in June. Transducers is the primary international conference for the MEMS field. Luckily for Manz, it was being held in nearby Tokyo that year. At the conference, he was enthused by the new range of MEMS devices that were being displayed. As soon as he returned to Hitachi, he proposed to build a miniature liquid analysis device inspired by the principles of the Stanford gas analyzer. Manz also knew about the work of Piet Bergveld, who had integrated chemical and biological sensors based on specially coated transistors inside silicon microchannels (figure 3.1).[1] Like Steve Terry's device, Manz's device would also have a planar "chromatography column"—a long channel, really—that he would use to separate complex mixtures into single compounds (figure 3.2). Unlike Terry, Manz would use his device to separate liquid mixtures, not gas mixtures. Making the column ten times as small would reduce the time it takes for molecules to diffuse across it—and hence speed up the separation—by a factor of a hundred: the benefits could be

Figure 3.1
Piet Bergveld's group at the University of Twente put coatings on transistors in 1985 to make them sensitive to chemical and biological molecules. The device was assembled inside a silicon microchannel (not shown).
Image courtesy of CBMS (Okaar Photography).

Figure 3.2
An array of liquid chromatography "columns" designed and fabricated by Andreas Manz in 1987.
Image courtesy of CBMS (Okaar Photography).

immense. Although the Miyagi lab did not have much microfabrication know-how, Manz quickly found that the group of Kazuo Sato in Tokyo could provide the needed expertise and the clean room. With Hitachi's approval, Manz started the project. A paper was eventually published in 1990 (after Manz had left).[2]

THE GENESIS OF MICROTAS

In 1988, Manz decided to leave Japan and look for jobs in Europe. One of these applications—at Ciba-Geigy in Basel (Switzerland)—would change the field of microfluidics forever. The Basel community at large was very concerned at the time with environmental safety because of the 1986 Sandoz chemical spill that had turned the Rhine red overnight and caused massive wildlife mortality downstream. Michael Widmer at Ciba-Geigy was interested in developing chemical monitoring systems and had recently written of "integrated, total analysis systems (TAS)."[3] He had postulated that instrument footprint and analysis time would decrease over time—but had made no mention of MEMS-based miniaturization. Manz got the job, and Widmer charged Manz with the vague task of building an online, fast, and highly selective chemical monitor that would be capable of detecting one hundred organic compounds. But otherwise—in a typical Widmer managerial style—he granted Manz ample freedom to pursue the project.

Andreas Manz sensed that microfluidics needed to formulate its fundamental scaling laws, a scaffold for the field to grow solidly. On March 8, 1989, he presented an internal memo[4]—typeset in white over blue background—at Ciba-Geigy (figure 3.3). In it, Manz proposed to build a miniaturized liquid chromatography and electrophoresis integrated chip (a "microTAS"). In this prescient document, Manz outlined all the benefits of microfluidics: the quadratic scaling laws ("10× smaller is 100× faster" due to diffusion), the smaller footprint, the savings in reagents, the sensor integration, the design modularity—the slides are a far-sighted road

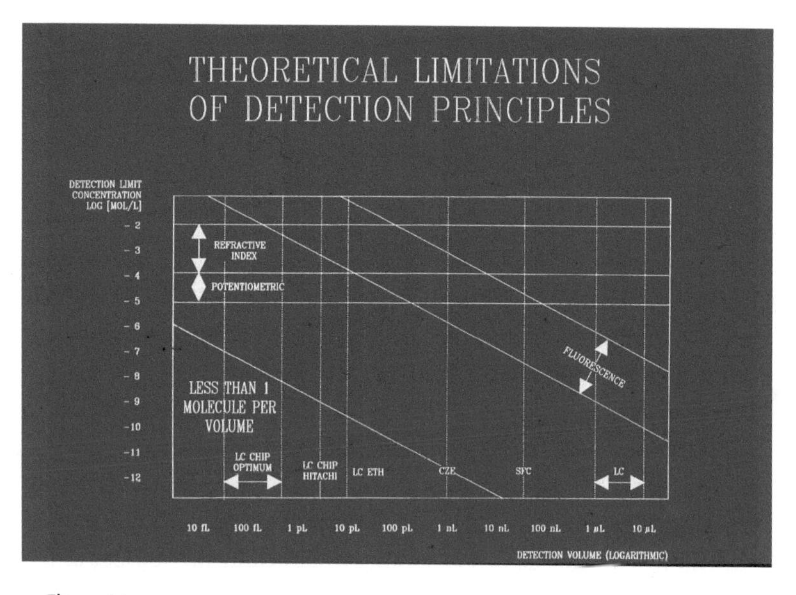

Figure 3.3
One of the slides of Andreas Manz's first presentation of the microTAS concept at Ciba-Geigy on March 8, 1989.
Image contributed by Andreas Manz.

map for the future of the field. Interestingly, the document does not contain the word *microfluidic*. It would be another few years before the word appeared in the literature for the first time.[5] Yet the document screams that the microfluidics field had just found in Manz its angel architect, just like MEMS had found an angel architect in Jim Angell a decade prior.

Manz's idea for integration of analytical systems in a miniaturized format resonated instantly with Widmer. The project was quickly approved, and in 1989 Manz designed the first chips in silicon. ICSensors, Steve Terry and Hal Jerman's company, fabricated the chips. The team finally published the work in 1992.[6] At the Transducers '89 conference in Switzerland, Manz met Jed Harrison, a chemistry professor at the University of Alberta who was working on integrated chemical sensors. "Together, we plotted a sabbatical leave for me [at Widmer's group], while he simultaneously hired Sabeth Verpoorte from my lab as a postdoctoral fellow," Harrison explained of that first encounter.[7] The group was well aware of parallel developments in Japan, Europe, and the United States in silicon. "Japanese researchers were developing integrated fluidic systems with biosensors at the same time,[8] while others in Europe[9] and the United States[10] were attempting pumps and valves in silicon, so the concepts were clearly beginning to germinate worldwide at that stage," remembers Harrison.[11] Soon Widmer's group switched to glass, fabricated with the help of Felix Schwager at Mettler-Toledo, for all its designs.[12]

The first glass microfluidic chip, published in 1992 in *Analytical Chemistry*, was a capillary electrophoresis chip built by Jed Harrison and Andreas Manz.[13] The chip also featured the first microfluidic valve—an innovative "electrokinetic valve" that was not able to stop the flow but could divert flow from one channel to another by switching the voltages of the inlets. In this device, the researchers decided to abandon the simplicity of chromatographic separations and instead use capillary electrophoresis by applying voltages with electrodes inserted at the inlets. About 3,000 volts were needed to run the separations, and 500-volt pulses to switch the flows. Thanks to the surface charges present on the glass walls, the fluid was pumped as a flat front (called a plug flow), which reduced lateral diffusion and temperature variations across the width of the channel. When they introduced a mixture of fluorescent molecules of different molecular weights at the inlet, the different molecules traveled down the channel at different rates, leading to separation, which could be visualized as bands on the channel with a fluorescence microscope. The only disadvantage of using electrical fields was that, for complex mixtures, the charge and the mass of the molecules could, in principle, compensate each other. In other words, a larger molecule that had twice the amount of charge could move as fast as a smaller molecule with half the amount of charge, but researchers soon learned to identify these problems based on the nature of the chemicals.

At some point, Manz, Widmer, and colleagues decided that this new field needed a formal introduction to society. They published their vision in a set of now-legendary papers where they formulated the physical, chemical, and engineering fundamentals of miniaturized chemical analysis systems.[14] The influential 1990 *Sensors and Actuators B* theoretical paper[15] by Manz et al. is now considered by many to be the foundational paper of microfluidics as a field, framing the road map for the developments ahead. As sketched in the Ciba-Geigy memo, the paper laid out the advantages of microfabrication: novel fluid phenomena leading to faster, more efficient analysis; batch fabrication and parallel operation of many channels or devices leading to lower costs; and the sensor integration ideas pioneered by Piet Bergveld. In their follow-up 1993 paper in the journal *Science*, Manz and coworkers presented a glass chip for high-resolution, high-speed electrophoresis.[16] This paper experimentally verified the predictions of the 1990 *Sensors and Actuators B* theoretical paper. Held since 1994, an international conference in microfluidics and biomedical MEMS aptly named MicroTAS is attended by thousands of scientists every year.

You can tell by looking at Andreas Manz that he has done well in life, and he now has time for moth-watching again. Microfluidic chips were the right career choice, after all. In 1995, together with Larry Bock, Michael Knapp, and Michael Ramsey, he founded Caliper Technologies (later Caliper Life Sciences). This company combined microfluidics, liquid handling, and laboratory automation to deliver research tools for drug discovery and development and for genomics and proteomics laboratories. In 2011, PerkinElmer bought Caliper Life Sciences for $600 million. To this day, he still drives hundreds of miles across Europe to conferences—for the pleasure of it—in his 1929 Bugatti or his 2006 Spyker. At the conference, you'll recognize him immediately because he is usually the guy sporting a black leather jacket, jeans, sunglasses, and a jovial gait among a sea of button-down-shirt engineers. Once he offered me a ride in his car and I had to cling to the door handle for extra support: you don't ask the daring architect of microfluidics to slow down.

THE LAUNCH OF GLASS CHIPS

Richard Mathies's interest in microfluidics started some years before the Human Genome Project (box 3.1). In the mid-1980s, while he was looking for a miniature refrigerator for his photoreceptor vision research, he stumbled upon MMR Technologies, a cryogenic company founded in 1980 by Stanford physics professor William A. Little. Many cryogenic devices require extreme cooling, from superconductors to quantum supercomputers. While at Stanford, Little had realized that a microfabricated cooler—which he termed microminiature refrigerator—would be more efficient due to its small size, low noise, fast response, and low cost.[17]

William Little's micro-refrigerator was far ahead of his time, so his ideas would need to wait several decades to have an impact. He could not have predicted that computers would rely on worldwide-web connectivity and cloud storage, both of which rely on massive data centers. Presently, data centers in the United States alone consume 24 terawatts-hour of electricity and 100 billion liters of water a year to refrigerate the computer mainframes. This amount of water and electricity is equivalent to a city the size of Philadelphia. The cost of the refrigeration adds more than 30 percent of the cost of running the data center. Recently, a team from the *École Polytechnique Fédérale de Lausanne* (*EPFL*) in Switzerland developed

BOX 3.1: BIOGRAPHY
Richard Mathies

Richard Mathies was born in Seattle in 1946. His father was an analytical biochemist who ran the clinical lab at Swedish Hospital for many years. Mathies liked building stuff from an early age and fabricated a simple X-ray powder camera while in high school. After a PhD in physical chemistry at Cornell University and postdoctoral research on the photochemistry of visual pigments with Lubert Stryer at Yale, he became a professor of chemistry at UC Berkeley in 1976. He stayed at UC Berkeley until the end of his career more than four decades later. It takes an entire life devoted to chemistry to decode the chemistry of life.

an integrated version of Little's chip that slashes this added cost to an insignificant 0.01 percent. They used silicon etching techniques to fabricate a subsurface network of buried microchannels a few microns below the active layer of a microelectronic chip.[18] By directly cooling a microelectronic chip with circulating water right from underneath through integrated microchannels, they could reduce the refrigeration costs by several orders of magnitude.

A predecessor of the EPFL's, Little's microminiature refrigerator was essentially a microfluidic device based on the thermodynamic effect known as the Joule-Thomson effect: as gases expand through a small orifice into a thermally insulated container, they must cool down. In Little's device, as opposed to previous cryogenic chambers, the size of all the components, from the chambers to the expansion tubes, had been reduced by a large factor, allowing for much faster cooling. Little designed the microfluidic device as a simple serpentine of planar capillaries that gradually increased in width, thereby causing the expansion and cooling of the gas. Photolithography and glass etching had not been developed, so Little came up with a fabrication process using his knowledge of eighteenth-century photographic processes. He defined the microchannel pattern by covering a glass substrate with silver gelatin—the same emulsion covering photographic paper; the silver halide salts in the gelatin make it photosensitive—exposing it to light and developing the pattern. The glass-exposed areas were etched using sandblasting with fine silica powder; finally, after removing the protecting gelatin pattern, the channels were bonded with an ultraviolet (UV)–curable glue. The devices were already being sold by MMR Technologies when Mathies first learned about them.

The microminiature refrigerator catalyzed Mathies's subsequent interest in the potential power to apply microfluidics to DNA analysis during the technology development phase of the Human Genome Project starting in 1987. Mathies argued that they had already been able to focus a laser on a small sample for their Raman spectra of photoreceptor pigments. Raman spectroscopy is a light-based chemistry technique that allows for identifying the structure of molecules and requires a challenging experimental setup. Therefore, it should be at least possible—if not straightforward—to do chemical analysis of DNA using microfluidics down to the single-molecule detection level.

Time proved his intuition right. In 1990, he hired Xiaohua Huang, a postdoc who had done his PhD in Richard Zare's lab at Stanford on capillary electrophoresis, to implement an array of glass capillaries. Huang read the fluorescence in the capillaries with a laser-excited, confocal-fluorescence scanner. Mathies passionately describes Huang as "the unsung hero of the Human Genome Project." Thanks to his work, published in 1992, the speed and throughput of DNA sequencing could be dramatically increased by performing sequencing separations in parallel on the capillary arrays followed by optical detection.[19] Yet none of these papers—written just as the first Manz papers on glass microfluidics were coming out—used microchip technology or the word *microfluidic*. That leap forward came from the work of Adam Woolley (box 3.2), a talented chemistry student that joined the Mathies lab in 1992 and who later would make many more seminal contributions to microfluidics (chapter 9).

When they first talked about a research project in 1992, Mathies proposed the idea of doing capillary electrophoresis DNA separations using the microfluidic technique just published by Manz. "That sounds like a one-year project—What would I do next?" inquired Woolley, who wanted to hear the professor's long-term vision. Mathies laid down a doctoral project on miniaturizing and integrating DNA sequencing processes in a microfabricated format, and Woolley started working on it right away.

The first problem to solve in that long PhD road was, indeed, the DNA separation by capillary electrophoresis using a microfluidic chip. It was not easy at the beginning because funding was tight. Woolley recalls having to share an essential piece of equipment for his detection setup. Only in his second year, after Woolley got some promising results, did Mathies decide they could spend money to get the dedicated equipment for Woolley's detection apparatus.

Finally, Woolley and Mathies were able to separate DNA fragments at ultrahigh speeds in glass microfluidic channels filled with a sieving matrix made of hydroxyethyl cellulose, a study published in 1997.[20] The

BOX 3.2: BIOGRAPHY
Adam Woolley

Adam Woolley has been surrounded by chemistry and the Mormon faith all his life. He was born in 1968 in Provo, back then a small town in central Utah where almost 90 percent of the population belonged to the Church of Jesus Christ of Latter-Day Saints, commonly known as the Mormon faith. A large concentration of Mormons lives in Utah because Brigham Young, one of the early leaders of the movement, settled there. Brigham Young founded Brigham Young University (BYU), Provo's private university that is still run according to Mormon principles. Mormons are characterized by strong family ties. Woolley's mother had a BS in chemistry, his father obtained a PhD in chemistry from BYU, his older brother also majored in chemistry, and his younger sister would become a chemical engineer. "I'd been interested in chemistry since a young age, and I had an excellent high school chemistry teacher [Marty Monson] who fostered that excitement for chemistry," recalls Woolley. Later his father became a faculty member of BYU's Chemistry Department. Thus, it was most natural for Adam Woolley to study chemistry at BYU. Because UC Berkeley has always had one of the top graduate programs in chemistry, "and I was attracted to its storied history of Nobel Laureates," Woolley decided to pursue his PhD at UC Berkeley. His initial interest was nuclear chemistry, "but after meeting Rich Mathies, I was especially intrigued at the prospects of miniaturization for DNA analysis," he explains. "I really enjoyed the years in Mathies's lab. I was on a paid fellowship, so he gave me broad latitude in my work," he adds.

work was the first and necessary step for microfluidic-based, high-speed DNA sequencing (published in 1995),[21] DNA genotyping[22] (the detection of small genetic variations or "genotypes" in a DNA sequence that can be considered a genetic fingerprint of an individual), and the integration of *polymerase chain reaction* (*PCR*) with a DNA separation chip,[23] all with the essential participation of Adam Woolley.

PCR is a method invented in 1984 by biochemist Kary Mullis to make many copies of—biologists like to say "to amplify"—a given DNA molecule until it reaches detectable levels (box 3.3). PCR, for which Mullis received a Nobel Prize in 1993, is now a common technique used in clinical and research laboratories for a wide variety of applications, including diagnostics, biomedical research, and criminal forensics.

BOX 3.3
DNA Amplification or PCR

PCR is a method for "amplifying" (i.e., making many copies of) a given DNA molecule. The method first uses heat to separate the DNA double helix into its two strands. Starting with a template that binds to one of the strands (the "primer"), a natural enzyme called DNA polymerase then rebuilds the double helix by adding the nucleotide bases (the encoding units or "letters" of DNA, which can be A, C, G, or T), one at a time, as a mirror image of the opposing strand, effectively building a copy of each strand much like a zipper closes up. Thus, each heat cycle results in the duplication of the number of DNA molecules. After, say, ten cycles, the number of molecules has multiplied by $2^{10}=1{,}024$, resulting in rapid, exponential amplification of DNA material.

But PCR needs heat cycling—heating the double-stranded DNA to separate it into two strands and cooling it down to allow for the DNA polymerase to build the double strands. The heat cycles are slow and cumbersome to control. One solution to this problem uses microfluidics. In 1998, Andrew deMello and Andreas Manz developed a clever continuous-flow DNA amplification chip.[24] Rather than heating and cooling down the whole chip, they designed serpentine-shaped microchannels that moved the reactants through heaters and coolers, which achieved world-record-speed DNA amplification without human intervention. The study was published in *Science*.

Another solution to the cumbersome PCR heat cycles is provided by LAMP (short for loop-mediated amplification reaction), a variation of PCR that can achieve amplification at a constant temperature of 60°C. Although less sensitive and less versatile than PCR, LAMP is faster and more straightforward so it is usually preferred for point-of-care, portable devices.

Because PCR had not been integrated as part of the device yet, the DNA samples often underwent an amplification step before being introduced into the chip. To combine PCR and capillary electrophoresis in a single microfluidic chip, Mathies started a collaboration with Allen Northrup at Lawrence Livermore National Labs, about an hour east of Berkeley. At a conference in 1993, Northrup et al. reported the first microfluidic chip for DNA amplification by PCR, which exploited the high thermal conductivity of silicon and small volumes of the chambers to speed up the thermal cycling reactions (figure 3.4).[25] This development led in subsequent years to a flurry of DNA analysis chips integrating more and more steps of the

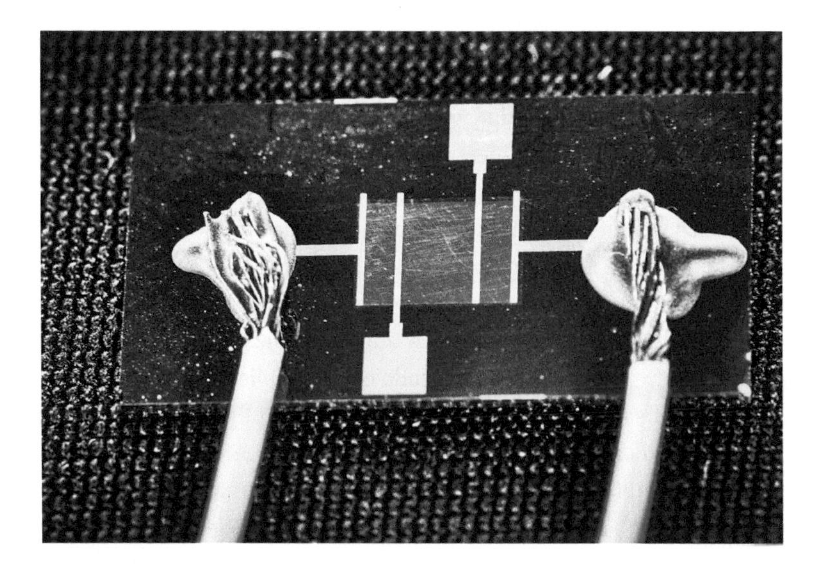

Figure 3.4
The base for the first microfluidic PCR device fabricated by Allen Northrup in 1993. The electrodes are used for heating. The microchannel (not shown) is assembled on top.
Image courtesy of CBMS (Okaar Photography).

DNA processing workload.[26] In the initial experiments, because Northrup wanted to have the experiments properly validated with his group's expertise, Adam Woolley would first drive the chip to Livermore, where Northrup's group would do the PCR. Then Woolley would put the device on ice in a Styrofoam box in his car, drive back to Berkeley, and continue with the capillary electrophoresis experiments on the same chip. "Once we got some decent results with that approach, Allen let us use his PCR setup at Berkeley, and that resulted in a highly cited paper in *Analytical Chemistry*" in 1996,[27] recalls Woolley. Mathies still remembers Woolley as "the brave pioneer who did the first microfluidics project at Berkeley."

Almost a quarter of a century later, microfluidic PCR chips are ubiquitous and commercially available, typically packaged in neat cassettes that interface with an electronic reader. During the COVID-19 pandemic, many companies (such as Fluidigm, Mobidiag, Quidel, Talis Biomedical, Mammoth Biosciences, and Mesa Biotech) have been able to rapidly deploy microfluidic PCR chips for the automated amplification of DNA to detect SARS-CoV-2 virus from nasal swabs or saliva.

But to enable sequencing of the 3.3 billion base pairs of DNA of the human genome, Mathies would need more than ultra-high-speed sequencing—he would need massive parallelization as well. After Woolley graduated with his PhD in 1997, Mathies continued to advance the project with his group, by now large and well funded. Eric Lagally integrated valves and temperature sensors for real-time control of the PCR reaction[28] and demonstrated single-molecule DNA amplification.[29] The speed of DNA separation by electrophoresis depends on the applied voltage. However, they were already operating at the limit: increasing the voltage a little bit higher would cause the sieving material they used inside the channels to overheat and ultimately fail. Therefore, the only option was to use massive parallelization, a strategy commonly used in the microelectronics industry to build more and more complex circuits. Peter Simpson joined the group and developed robust methods for manufacturing high-density, micron-precision devices with high-quality channels and defect-free surfaces. The group was initially able to operate 6-inch wafers containing 96 electrophoresis channels (figure 3.5),[30] and later they upscaled to 8-inch wafers containing 384 channels (figure 3.6)[31] "that remain to date a record for the field," Mathies proudly points out.

This research ultimately led to the commercialization of the MegaBace DNA sequencer. This sequencer and the Perkin-Elmer 3700, based on the capillary-based DNA sequencing research by Norm Dovichi's group at the University of Alberta in Canada, were the two primary DNA sequencing machines used by groups spread across the globe to complete the human genome. In the end, a legion of microfluidic devices helped to decode the human genome.

ULTRASENSITIVE DNA ANALYZERS

Microfluidic DNA analysis has found many other uses, and Richard Mathies has always been restless. "I like to start a whole new thing every

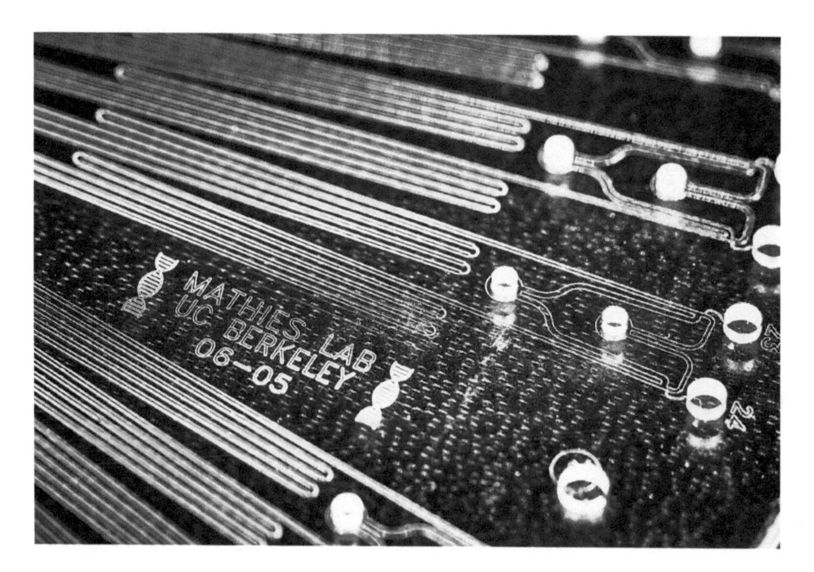

Figure 3.5
A detail of a microfluidic DNA sequencer based on capillary electrophoresis capable of ninety-six parallel measurements. The device is a large circular disc—only the channels corresponding to two or three measurements are shown here. The meandering channel design is used to increase the DNA separation efficiency. A cathode for applying voltage is introduced in the middle small hole. The larger holes are for introduction of the sample. At the time (2002), this instrument was already five times faster than competing non-microfluidic capillary electrophoresis platforms.
Image courtesy of CBMS (Okaar Photography).

ten years," he says. NASA was interested in analyzing biomolecules found in space, which led to NASA funding his work on using microchips to separate amino acids starting in 1997. Mathies saw a broader opportunity in space: the application of microfluidic sample preparation, processing, and analysis for the chemical exploration of our solar system. It is believed that for life to start on Earth and elsewhere, the molecular building blocks of life need to be of predominantly uniform "handedness." More complex molecules such as proteins and enzymes would not work properly if some of these blocks were mirror images or chiral versions of

Figure 3.6a
A 384-channel DNA sequencer; 384 pairs of electrodes are introduced at the inner and outer circles to apply the voltage needed for capillary electrophoresis. As the device rotates, fluorescence readouts can be performed on the outer circle using a custom-designed microscope.
Image courtesy of CBMS (Okaar Photography).

Figure 3.6b
Detail of figure 3.6(a).
Image courtesy of CBMS (Okaar Photography).

each other.[32] In 1999, inspired by this hypothesis, Mathies and his PhD student Lester Hutt built a microfluidic device that could electrophoretically separate amino acids with such precision that it could distinguish an amino acid from its mirror-image molecule.[33] That means that the microfluidic analyzer by Mathies and Hutt produced, for every amino acid of biological origin, two adjacent peaks: one for the "D" or right-handed version and one for the "L" or left-handed version. If the amino acid was not of biological origin, then only one peak was present. Another PhD student, Alison Skelley, joined the lab in 2001 to continue Hutt's efforts.[34] Canadian-born, Skelley had already been introduced to chemical separations by capillary electrophoresis during her undergraduate research with Norm Dovichi at the University of Alberta. In the Mathies lab, she started the long development of the Mars Organic Analyzer (MOA), a complex microfluidic instrument initially scheduled to launch on the Mars Rover.

The core of the MOA was a four-layer microfluidic capillary electrophoresis chip for the analysis of amino acids present in the soil of Mars.[35] The chip contained pneumatic microvalves (i.e., actuated by pressurized air) and an optical detector, which required significant instrumentation around the chip. "We were fortunate to have a very skilled engineer in the lab, James Scherer, who built that box," Skelley remembers. The affectionate reference to "that box" alludes to the complexity within it, besides the microfluidic device itself (figure 3.7). The first time Scherer and Skelley took the MOA for a field test, it was to the Golden Gate Bridge 20 miles west of campus. "I remember sitting in the Marin Headlands overlooking the bridge on a picnic bench, running the MOA from a car plug adapter," she says. The MOA required electricity to run the optical detector and the vacuum pump for the valves, hence the need to be connected to the car battery. "Sitting there with a black box with colorful wires coming out (being an international student no less)—probably I wouldn't be able to get away with that now without some serious questions!" she jokes.

Once Skelley had shown that the MOA could work in an urban setting across the bay, Mathies wanted to test it in an environment as similar to Mars as possible. He sent the MOA with Skelley and colleagues from a

Figure 3.7
Alison Skelley setting up the Mars Organic Analyzer on a park overlooking the Golden Gate Bridge.
Image courtesy of Alison Skelley.

very large team, including the Jet Propulsion Laboratory and the NASA Astrobiology Institute, to the Atacama Desert in Chile. With 1 millimeter of rainwater per year, the Atacama Desert is the driest place on Earth. Death Valley, by comparison, is flooded with several inches per year. Mathies describes the Atacama as "like a Martian landscape—no twigs, no leaves." They set up camp in the Atacama for a couple of weeks, testing various instruments and collecting samples. They used a white trailer as a portable, albeit cramped, laboratory (figure 3.8). "It was an amazing experience," says Skelley, who also uses other brushstrokes to describe the expedition, such as "exhausting" and a "stressful ordeal." Mathies joined for a second trip. The goal of the study was to see if they could detect any signs of life. "Probably one of the toughest tests for a biodetection assay," emphasizes Mathies.

Figure 3.8
(From left to right) Pascale Ehrenfreund, Frank Grunthaner, and Alison Skelley posing with a later version of the Mars Organic Analyzer inside the trailer in the Atacama Desert.
Image courtesy of Alison Skelley.

The soil of the Atacama Desert was indeed arid and harsh (figure 3.9). Acids and dust deposited in the soil form a "duracrust" surface acidic layer capable of modifying organic molecules, completely obliterating their chemical bonds under intense sunlight. On the first samples taken from the surface, the team did not detect any amino acid signatures when they ran the analyzer at night inside the trailer. Skelley then took another set of samples from soil shielded by rocks and dug 2–5 centimeters to obtain subsurface soil. This second set of samples showed an intense signature of both D and L amino acids. She could pinpoint these signatures to a ten- or hundred-thousand-year-old biological origin thanks to the high sensitivity of MOA.[36] As similar conditions likely apply to the dried river

Figure 3.9
Richard Mathies in the Atacama Desert.
Image courtesy of Alison Skelley.

formations seen on Mars, researchers hope to use the MOA or a similar microfluidic DNA analyzer to analyze Martian samples.

The box hides the sophistication of the MOA. Initially, valve integration enabled a hands-off approach to run multiple samples on a single device and prepare samples, such as mixing soil extract with dye. But traditional microfluidic devices are specific to the assay (the application) they need to run—they are not programmable. Sending a device to Mars meant that NASA might need to deal with unforeseen circumstances and reprogram the device on the spot. Skelley did not know how to do that, but she was lucky to find someone else in the lab—William Grover—who did (box 3.4).

Grover was running the email server for the Mathies group when Alison Skelley arrived. She was introduced to him for help to set up an email account. "I vividly remember thinking, 'wow, she's beautiful' and feeling nervous but trying to act all professional while setting up her email."

BOX 3.4: BIOGRAPHY
Will Grover

Will Grover's lifelong interest in medical technology naturally arose from his early experiences in Johnson City, Tennessee, a small town in the foothills of the Smoky Mountains. Born in 1977, he volunteered as a kid in a large Veterans Affairs hospital where his dad worked as a nurse in an operating room. He remembers "being fascinated by all the medical equipment the surgeons and nurses used to diagnose and treat the veterans." His dad would snag unused and discarded plastic tubing fittings for intravenous lines and the like for him to play with. "I loved connecting together the little tubes and valves and whatnot and watching water flow through them," he recalls vividly. During high school, he participated in summer educational programs at the University of Tennessee in Knoxville (where he ended up studying chemistry as an undergrad) and at the University of California, Berkeley and the nearby Lawrence Berkeley Laboratory (where he ran his first PCR and gel electrophoresis). Grover credits these high school experiences for sparking his career in science. He fell in love with the city of Berkeley and the university and resolved to return one day. He achieved that goal by doing his PhD at UC Berkeley's chemistry department under Mathies's supervision.

When the lab moved locations, Mathies put both Skelley's and Grover's projects in the same tiny windowless room. "The ceiling leaked right over our shared optical table whenever it rained, so we were huddled in this small room for hours every day," explains Grover. Skelley and Grover began a very prolific collaboration and a lasting friendship.

Grover is extremely creative and had then one of his brilliant ideas, which proved very helpful to Skelley's project and would influence many others. Grover had always been fascinated by nonstandard modes of computing—those that do not use electronics. For his original project, he had discussed with Mathies making a microfluidic "computer" that would use DNA to perform basic logical operations. Instead, he developed a different type of unconventional computing device based on fluidic logic.

Grover, of course, knew about the MONIAC, an analog cupboard-sized computer that used simple fluidic logic, created in 1949 by New

Zealand economist Bill Phillips to model an economy. As a kid, he used to love watching the pressurized-air (pneumatic) tubes for passing messages around at the hardware store. "Discovery is seeing what everybody else has seen, and thinking what nobody else has thought," as attributed to Dr. Albert Szent-Györgyi, the Hungarian biochemist and 1937 Nobel Prize winner. One day, Grover looked at existing microfluidic pneumatic valves—essentially fluidic switches using pressurized air (chapter 5)—and thought of a transistor—an electronic switch, but conceptually equivalent to a valve. Therefore, he reasoned, he could design pneumatic digital logic gates (AND, OR, NOT, NAND, and XOR) of the same type that underlie the operation of integrated microelectronic circuits, using networks of pneumatically actuated microvalves instead of transistors. "We could immediately start designing more complex pneumatic 'circuits' by copying electronic circuits, and by and large, they just worked as we expected. It was probably the only research project in my life that 'just worked' the first time," Grover recalls. They first demonstrated a microfluidic pneumatic device that added 8 bits.[37] Later, building on Grover's framework, others developed flip-flop circuits,[38] bistable oscillators,[39] a clock oscillator with suitable timing accuracy to control diagnostic assays,[40] and an autonomous octopus-shaped robot.[41] An attractive feature of such microfluidic logic circuits is that they are immune to electromagnetic interference, so they would not be corrupted by highly ionizing radiation in space. For the MOA, Skelley added fluid paths and failure modes to support additional operations so the chip might reconfigure itself and recover, should the need arise on Mars, without the hands-on help of a human (since the nearest human might be more than 50 million kilometers away).

The MOA was not selected by NASA to go to Mars inside the Perseverance Rover that blasted off in July 2020. In the end, on this rover NASA opted instead for seven remote optical analyzers—such as lasers, cameras, and radars—possibly because of unresolved risks of robotic sample loading in microfluidics. The MOA did feature an unprecedented parts-per-trillion sensitivity, enough to detect subsoil amino acid concentrations in extreme environments on Earth—and astronauts could use it in future missions. On our planet, it will be remembered as the first autonomous microfluidic analyzer—as well as the catalyst for Alison Skelley and William Grover's marriage in 2006.

* * *

Mathies, who is now an emeritus professor, did not stop there. He started IntegenX, a company that commercialized his microfluidic DNA analysis technology for forensics and could help investigators gather reliable evidence to convict or exonerate suspects of crimes and reconstruct what happened in a catastrophe. The tabletop device uses integrated PCR and capillary electrophoresis to identify sets of short tandem repeat markers in a DNA sample that provide a unique fingerprint of a person in one hour. The microfluidic device is in a cassette, hidden from view, of course. The strength of this forensic approach is that researchers can apply it to the DNA of any species, not only humans. DNA found in fossils and ancient remains has helped scientists draw a more accurate picture of the evolution of life on Earth. A pathologist may use DNA sequencing to analyze which microorganisms are growing in a hospital and reduce the risk of creating antimicrobial resistance. Virologists routinely identify viruses, which are too small to be seen under a microscope, by sequencing their DNA. Scientists use viral sequencing to determine the origins of an outbreak during an epidemic such as COVID-19.

Mathies's current projects look for chiral amino acids as chemical signatures of life on the icy moons of Saturn and Jupiter and in the tails of comets where autonomous processing and analysis in an extreme environment is most important. Richard Mathies should be proud of himself: not only did his microfluidic chips help decode the human genome, but NASA is also still considering launching them toward another planet to search for traces of extraterrestrial life. As Woolley and Mathies like to say, "The sky is not the limit for microfluidic systems."

4 MINIATURIZED TESTING WITH THE GLUCOMETER AND OTHER POINT-OF-CARE DEVICES

We have all been to the doctor to get a blood test. If your body can't handle sugar or cholesterol or you have an infection, it shows up in the test and the doctor can prescribe a treatment. These tests are performed with microfluidic blood analyzers. Portable microfluidic instruments have revolutionized the ability of modern medicine to provide diagnostics right where the patient is—at the so-called *point of care*. From people with diabetes testing their blood sugar to better manage their disease to rapid blood analysis performed in the clinic or hospital, these devices have silently played their role. We almost take for granted that they have provided near-instantaneous health monitoring with minimal patient discomfort.

The fifty years that have elapsed since the late 1960s when my friend Jeff Spencer was diagnosed with type 1 diabetes have seen a revolution in glucose sensors. As a child, when he learned he had diabetes, his doctor issued him a cruel warning: "You are going to go blind, and you are going to lose your feet." His first home monitoring of blood sugar in the 1970s relied on color-changing paper strips. Now he wears the latest technology: a continuous electronic glucose sensor that "speaks" to intelligent insulin pumps through Bluetooth. The heart of these revolutionary improvements in health care has been microfluidics, enabling not just miniature glucose sensors but also many other point-of-care devices.

THE SWEET URINE AND BLOOD OF DIABETES

Microfluidic glucometers developed at the beginning of the twenty-first century enabled the frequent analysis of tiny amounts of blood (figure 4.1) and have dramatically improved the lives of millions of people. These glucometers have eased the self-monitoring of glucose in the blood—avoiding

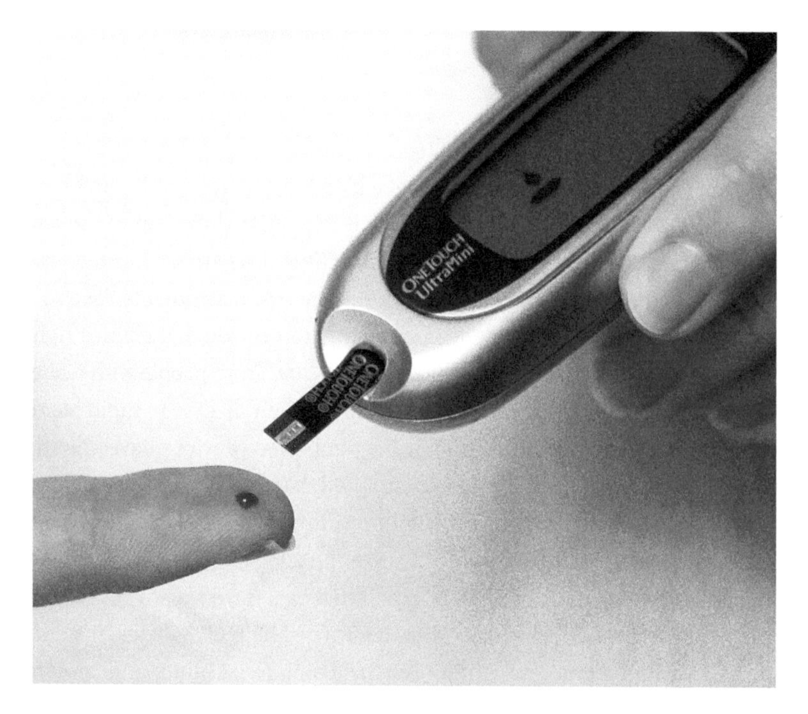

Figure 4.1
Microfluidic glucometer. Upon touching the finger, the blood drop wicks into the strip by capillarity. Inside the strip, electrodes loaded with an enzyme perform an electrochemical reaction in a few seconds and report the result to the handheld electronic reader. The strip is disposable.
Image source: Pxhere.com.

life-threatening episodes of hyperglycemia or hypoglycemia (too much or too little sugar in the blood)—so patients are now better able to take responsibility for their lives. But it was not always like this.

The care of people with diabetes (box 4.1) has a long history. The Egyptians knew about the disease in 1550 BCE and used the attraction of ants to the sweetness of a diabetic's urine to diagnose it.[1] In 1675 CE, the English physician Thomas Willis added the Latin term "mellitus" or "honeyed" to diabetes after tasting the sweetness of diabetic urine, which people knew about for centuries. The transition to a less subjective approach to the analysis of diabetic urine began with the British physiologist Matthew Dobson in 1776. He drew blood from diabetic patients, separated

BOX 4.1
Diabetes Mellitus

Diabetes—technically termed diabetes mellitus—is a severe and widespread metabolic disorder characterized by the body's inability to adjust the levels of glucose in the blood. The organ responsible for glucose regulation and balance is the pancreas: when blood glucose is too high, the pancreas secretes more insulin—a hormone that tells your cells to absorb glucose from the blood; when blood glucose drops, the pancreas releases another hormone, glucagon—which tells the liver to release its stored glucose into the blood. Glucose is necessary to provide energy to the cells, but high blood glucose levels can damage our bodies. In diabetic patients, the pancreas either does not make enough insulin (type 1 diabetes), or the body cannot normally respond to the insulin made by the pancreas (type 2 diabetes). Both scenarios cause blood glucose levels to rise.

Diabetes can cause life-threatening complications if left untreated. These complications include coronary heart disease, stroke, chronic kidney disease, foot ulcers, and damage to the nervous system and eyes. The disease affects about 400 million people worldwide, a number that has been increasing in the last decades due to changes in lifestyle. The incidence of diabetes has nearly doubled in the last forty years: in 2017, around 9 percent of adults worldwide had diabetes, compared to 4.7 percent in 1980. Thus, there has been a lot of interest for quite some time in point-of-care methods and instruments that monitor blood glucose, so-called glucometers. The glucometer can communicate the information to an insulin pump that artificially restores insulin to keep blood glucose levels normal.

plasma by sedimentation of the blood cells, and reported that, like the urine, the plasma also had a sweet taste. Dobson also isolated brown crystals from the dried urine, identified them as sugars, and found them to taste like brown sugar.[2] Less than a century later, in 1850, a Parisian chemist named Edme Jules Maumené created a simple method to detect high sugar in urine using a chemical reaction. He took a strip of sheep wool, impregnated it with tin chloride, and then soaked the strip with the patient's urine. If it had high sugar content, the strip would turn black upon heating (box 4.2).[3]

A more direct and straightforward approach using paper—another microfluidic substrate (chapter 7)—was designed in 1883 by a British doctor, George Oliver. He used reagent papers for testing urine at the bedside

BOX 4.2
Redox Reactions

Maumené's method used a type of chemical reaction called redox (short for reduction-oxidation) reactions. When two molecules react by exchanging an electron, chemists say that the molecule that has gained an electron has been oxidized, and the molecule that has lost an electron has been reduced. We can imagine the process as a handshake: the oxidizing agent agrees to give an electron to the reducing agent, and only to that molecule. In some cases, as in Maumené's, heat is necessary to bring the two molecules to a point where they can agree to make the handshake happen. In Maumené's reaction, sugar(s) in the urine was the reducing agent being oxidized, and tin chloride was the oxidizing agent being reduced. To be more precise, the tin chloride impregnated on the wool strip reacted with a specific chemical group called an aldehyde group present in sugars such as glucose. Glucose is a molecule formed of a ring of six carbon atoms with an aldehyde group tail. Upon heating, the aldehyde group in the glucose of the urine of Maumené's patients became oxidized and reduced the tin chloride to metallic tin, which has a black color. This chromogenic (color-generating) reaction had two drawbacks. First, other sugars such as fructose also contain an aldehyde group, so their oxidation reaction was not specific enough to signal the presence of glucose alone. And second, it was not quantitative; in other words, the coloration could not easily be correlated with the concentration of sugar in the urine.

using the reduction of alkaline indigo carmine to detect sugar with a simple, color-changing chemical reaction. The change in color, however, did not provide a measurement of the concentration of glucose.[4] These chemical detection methods also suffered from non-specificity: they could cross-react with other reducing substances present in the urine, such as fructose, galactose, uric acid, ascorbic acid, ketone bodies, and salicylates. Nevertheless, they persisted until the introduction in the 1950s of methods based on enzymes—molecules that recognize and specifically react with glucose.[5]

In 1957, the Ames Corporation developed the Clinistix, the first of the urine reagent strips that specifically detected glucose based on the oxidation of glucose by the enzyme glucose oxidase.[6] That same year, Joachim Kohn, a British pathologist, reported in the prestigious medical journal the *Lancet* that one could use the Clinistix to detect glucose from a drop of blood.[7] This critical advance dramatically minimized the amount of blood needed per patient, making frequent blood glucose monitoring feasible and opening the path to self-testing of blood glucose in the future. Ames Corporation then launched the Dextrostix, a platform that included strips and an optical reader to measure sugar in the blood. "As a boy [in the 1970s], I remember peeing on a paper strip, and when it turned green, my mom had me run around the block to burn the extra carbs—that's all we had back then," remembers Jeff. Various other strips impregnated with glucose oxidase, both for urine and blood, were developed and used well into the 1990s, along with readers that interpreted the result of the enzymatic reaction. Still, the measurement—as is often the case with optical readouts in non-laboratory settings—was susceptible to human error.[8]

The first instruments with direct electrical readouts used large electrodes that produced an electrochemical reaction with the blood (box 4.3). These were not portable, and they required too many operator-dependent steps. Initially, they were neither precise nor user-friendly enough to be left in the home of patients for them to interpret the results. These first blood glucometers were in hospitals, which only provided infrequent

BOX 4.3
The Clark Electrode and the First Glucose Sensor

The development of sensors with direct electrochemical transduction provided objective and quantitative glucose measurements. The first electrochemical biosensors that could deliver an electrical readout of the amount of glucose in the blood were developed in 1962 by Leland Clark and Champ Lyons at Cincinnati Children's Hospital. Earlier that same year, Clark had invented the oxygen sensor that bears his name. The Clark electrode measures oxygen dissolved in fluids such as water and blood using a platinum electrode covered with a semipermeable membrane that only lets oxygen (but not water) through. Platinum is used both as a catalyst—a surface that activates the chemical reaction without participating in it—and as a metal to supply electrons for the reaction. At the surface, oxygen gas and protons dissolved in the liquid react with electrons to form water molecules. The number of electrons that pass per unit of time is the electric current that, under ideal conditions, depends on the concentration of oxygen present in the solution. Although this reaction works for a while without the semipermeable membrane, Clark added it to prevent electrochemical deposition of impurities on the platinum electrode and thus lengthen the life of the device.

Clark and Lyons's glucose sensor was composed of a counter-electrode in addition to a Clark electrode and the semipermeable membrane was impregnated with the enzyme glucose oxidase. As glucose diffused through the membrane, glucose oxidase broke it down and produced hydrogen peroxide (and other reactants). Finally, hydrogen peroxide diffused to the electrode, where it was oxidized into oxygen, hydrogen, and one electron—producing an electrical current that was proportional to the initial concentration of glucose. For this invention, Clark is considered the father of biosensors. Every glucometer developed since relies on this reaction or a small variation of it.

Researchers have improved Clark's glucose sensor by immobilizing the glucose oxidase in hydrogels[9]—cross-linked, insoluble polymer networks containing more than 90 percent water. The water in the hydrogel helps maintain glucose oxidase in a more physiological and active state. Another substantial improvement has been the introduction of redox mediators (such as ferrocene) that bypass the need for relying on hydrogen peroxide. These redox mediators make better sensors because they do not react with oxygen, are stable in both the oxidized and reduced forms, are independent of pH, and react fast and reversibly.[10]

blood analyses for the average diabetic. My friend Jeff, who has followed with keen interest the developments of glucose monitoring technology, recalls with terror his biannual glucose monitoring visits in the 1970s at the Milwaukee County Children's Hospital near Lake Michigan. "Twice a year, three times in a row the day of the scheduled visit, all ten of us diabetic kids were lined up and watched the next kid being punctured in the fingertip with a sharp blade that felt like a can opener on the skin, and they took the blood into the lab," describes Jeff. "It looked like a camping accident."

Unfortunately, health-care providers believed then that blood glucometers could not be operated at home by patients. That is, until an engineer and diabetic patient named Richard Bernstein proved them wrong.[11]

Bernstein suffered from type 1 diabetes and was on insulin treatment. Whenever he missed a treatment, he would have a hypoglycemic attack, and he was starting to develop other complications of diabetes. Fortunately, Bernstein's wife was a doctor and was able to get him a machine, which he used to monitor his blood sugar and adjust his insulin accordingly. As a result of his self-monitoring, Bernstein's frequent hypoglycemic attacks were resolved. His health improved.

Although the procedure worked for Bernstein, it was not user-friendly. As Jeff recalls his own experiences with the machine once it became more widely available, "It was the size of a brick. I used it for a month twice a day and then stopped using it." The test was a cumbersome five-minute process involving a calibration step and a critical, harrowing ten-second delay, among other steps.

Bernstein personally convinced Ames Corporation to try to market the machine to patients, but the journals refused to publish his studies. Bernstein was so determined to prove that self-monitoring of blood glucose worked that, in his forties, he entered medical school, specialized in endocrinology, and started his own diabetic clinic. He performed and published the first studies in the 1980s with patients from his clinic and started the trend of allowing patients to monitor their sugar levels.[12] In the 1990s, manufacturers finally became interested in offering smaller, more reliable, and more user-friendly equipment—including a lancing device to facilitate the extraction of a blood drop.

THE MICROFLUIDIC ELECTROCHEMICAL GLUCOSE METER

In the twenty-first century, glucose monitoring microfluidic electrochemical sensors for diabetes management have been a blessing for diabetic patients. These sensors have truly brought patient self-monitoring to reality. The OneTouch Ultra was introduced in 2001, and within two or three years similar devices filled the market—MediSense's SoftSense, TheraSense's FreeStyle, OneTouch's UltraMini, and many others—requiring 1 microliter or less of blood. The FreeStyle, invented by Adam Heller, required only 0.3 microliters of blood (box 4.4). It also used a more reliable enzyme chemistry and type of electrochemical detection. These improvements made it more resistant to changes in temperature or hematocrit (blood cellular density) and unaffected by high concentrations of paracetamol (the generic name for the painkiller Tylenol), uric acid, and vitamin C.[13]

Adam Heller had developed a novel technology that allowed him to connect enzymes to electrodes using hydrogels, avoiding the mediation of a diffusing molecule (box 4.5), so he decided to team up in 1994 with his son Ephraim to commercializing a continuous glucose monitor like the one that my friend Jeff wears today. Ephraim had studied theoretical physics at Harvard but was studying at the Yale School of Management. They had the idea that, by continuously monitoring glucose, a wirelessly connected insulin pump could, in principle, adjust the necessary insulin levels as the patient eats normally. At the time, glucose monitoring systems already came with a lancet to prick one's finger. One day, experimenting with a lancet, Ephraim noticed that by pricking his finger, he could obtain a large drop (~5 microliters) of blood—with pain. However, if he pricked his arm instead, he obtained a blood sample that was ten times smaller (~0.5 microliters)—but *painlessly*. "Can we make a sensor for such a small sample of blood?" Ephraim quickly asked his father

BOX 4.4: BIOGRAPHY
Adam Heller

When the young boy Adam Heller was saved from certain death at Auschwitz, nobody knew that he would contribute to improving the lives of millions of human beings more than fifty years later. Heller was born in Romania in 1933 in Cluj, the vibrant capital of the region of Transylvania with an ancient and turbulent history. He was only nine years old when the Nazi-aligned Hungarians sent his father to be a slave laborer in Ukraine. In 1944, Heller and his family, along with eighteen thousand other Jews, were deprived of all their property and forcefully relocated to a ghetto. Although most of these Jews ended up being deported to the extermination camp Auschwitz, the Heller family was spared on the Kasztner train, which in July 1944 carried more than 1,600 Jews to the German concentration camp of Bergen-Belsen. Journalist Rudolf Kasztner had negotiated with senior German SS officers Adolf Eichmann, in charge of the deportations to Auschwitz, and Kurt Becher, in charge of the looting of Jewish property in Hungary, to allow some Jews to escape the gas chambers in exchange for money, gold, and diamonds. After six months, in December 1944, when the Third Reich was already in the final stage of collapse, they were released to safety in Switzerland. Adam Heller was then eleven years old.

Heller began his education in a kibbutz after emigrating to Palestine (now Israel) in 1945. When the time came to serve in the military, his interest in medicine led him to work in the pathology institute of a military hospital. Yet the lack of rigor of the medical science disappointed him. Heller then studied at the Hebrew University of Jerusalem, where he received an MSc and a PhD in physical organic chemistry. He left Israel for the United States, where he worked at UC Berkeley, Bell Labs in New Jersey, GTE, and the University of Texas at Austin.

and his father's colleague Ben Feldman. They answered that they could, because thanks to the microchannel's tiny volume, the electrooxidation of glucose would be fast and the total current measured (until the current dropped to zero when there was no more glucose to oxidize) would be proportional to the initial concentration of glucose. Adam and Ephraim Heller founded TheraSense in 1996 to develop FreeStyle, a device that realized that vision.

TheraSense released the FreeStyle in 2000. It was the first mass-manufactured fluidic device that required less than 1 microliter of fluid

BOX 4.5
Heller's Electrically Wired Redox Enzymes

An *enzyme* is a biological molecule that can catalyze or speed up a reaction between two biomolecules. *Redox reactions* (short for reduction-oxidation reactions) are a class of chemical reactions where a molecule exchanges an electron with another reacting molecule (box 4.2). Chemists say that the molecule that lost the electron became oxidized and the molecule that gained the electron became reduced. An example of a *redox enzyme* is glucose oxidase, the enzyme that catalyzes the oxidation of glucose.

At Bell Labs, around 1987, Adam Heller conceived the pioneering concept of electrical wiring of redox enzymes—the radical notion that one could connect redox enzymes to electronic circuits without diffusing redox mediators (box 4.3). To develop that idea, he joined the Department of Chemical Engineering at the University of Texas at Austin in 1988. Between 1987 and 1994, Heller's group succeeded in making the electrical connection using redox hydrogels,[14] where the flux of electrons was by electron transfer between mobile, yet polymer backbone-bound, segments. These hydrogels constitute the only known electron-conducting aqueous-phase materials, in which biochemicals like glucose and enzymes dissolve. "We coated electrodes with the enzyme-loaded electron-conducting hydrogels, creating technology for the transduction of biochemical fluxes to electrical currents," explains Heller. "We did perhaps the most important thing that helped people." With the redox hydrogels, he and his team developed and established the feasibility of miniature, subcutaneously implanted continuous glucose monitoring systems, based on the electrical wiring of glucose oxidase.[15] To reduce the time of the assay from forty-five to ten seconds, they switched to a quickly dissolving redox mediator—also based, like the redox hydrogels, on an osmium ion ($Os^{2+/3+}$) complex.

(blood). This product can monitor the glucose concentration in 300 nanoliters of blood,[16] "so little (1/8th of a typical blood meal of a mosquito)," clarifies Adam Heller proudly, "that for the first time diabetic people could sample their blood without pain." The device works by spontaneous capillary filling when the patient places a drop of blood at the tip of the strip, which acts as the inlet (figure 4.2). The FreeStyle devices are machine-printed on large rolls of polyethylene terephthalate (PET). "PET is inexpensive, and adequately maintains critical dimensions of devices

Figure 4.2a

FreeStyle strips for blood glucose monitoring. As a blood drop contacts the tip of the strip, blood enters the strip by capillarity, filling the space above the electrodes.

Image by Albert Folch, University of Washington.

Figure 4.2b

At this point, the strip is ready for a reading by the electronic glucometer, which will provide the readout.

Image by Albert Folch, University of Washington.

printed on it," explains Heller, giving credit to his colleagues Ben Feldman and Phil Plante for the fabrication. The electrodes are mass printed on two facing PET sheets that are precisely aligned, adhesive-bonded, and cut to produce devices with facing electrodes separated by precisely 50 microns. A surfactant is applied to the channel to accelerate filling. "Without the surfactant, filling the PET channel is slower than the few-second-long glucose oxidation current itself. The assay takes only seconds because the diffusion length is less than the 50-micron gap between the facing electrodes," points out Heller. In 2004, just four years after the launch of FreeStyle, Abbott Laboratories acquired TheraSense for $1.1 billion, which became the major part of today's Abbott Diabetes Care. Heller's colleagues at Abbott subsequently pursued and accomplished the original objective of a continuous glucose monitoring system based on a wired glucose oxidase. Released in 2014, the FreeStyle Libre is essentially a skin patch with a tiny transdermal microneedle that—without blood sampling for calibration—monitors glucose for two weeks and reports the results wirelessly.

My friend Jeff wears a similar microneedle skin patch from OneTouch that connects to his insulin pump via Bluetooth. He can safely eat a bagel after dialing 65 units of carbs so that his insulin pump anticipates his blood glucose rise. Thanks to an automatic regulation mode, the line on the screen of his handheld monitor that displays his glucose levels, updated every few minutes, stays relatively flat—as it would be in a non-diabetic patient—and he no longer sees wild fluctuations anymore.

These devices and improvements are a dream come true for diabetes management. And they proved the heartless predictions of Jeff's childhood doctor wrong: he is not blind, and he still had both feet last time he lounged on my sofa to watch TV.

IMANTS LAUKS AND THE I-STAT

Blood is our most vital fluid. It delivers nutrients and oxygen to cells and transports metabolic waste products away from those same cells. Cells, proteins, and hormones present in blood roam our bodies to regulate our well-being, always fighting invaders, adjusting our temperature, thirst, and even sexual appetite. Therefore, engineers have focused much effort in designing apparatuses that analyze multiple parameters from blood, not just glucose or cholesterol, for both clinic and home use. The advent of commercial microfluidic blood analyzers stands as one of the most fascinating pursuits in the history of health-care technology.

The first microfluidic blood analyzer (and the first microfluidic health-care product) was the *i-STAT*,[17] the highly successful device invented by British chemist-turned-engineer and entrepreneur Imants Lauks (box 4.6) in 1984. If you have ever been to a hospital and the nurse or the doctor came back with the results of a blood analysis within ten minutes of drawing your blood, it's likely that an i-STAT (or similar device) analyzed your blood.

Lauks realized that, for monitoring tests, the value of a result rapidly diminishes with the time that has elapsed since the doctor has taken the test. Usually, large hospitals are able to maintain a central laboratory with highly trained professionals and precision instruments that take high-accuracy measurements. In principle, this approach represents an efficient use of labor and reduces the cost of reagents. However, for urgent samples, these advantages are mitigated by delays in transport, queuing

BOX 4.6: BIOGRAPHY
Imants Lauks

Imants Lauks achieved early success on the standard academic path before taking the risk to leave it behind and pursue his dream of a rapid blood analyzer. Born in the UK in 1953, he earned an undergraduate degree in chemistry and a PhD in electrical engineering, both at the Imperial College London. He emigrated to the United States in the 1970s. By the time he was only thirty years old and decided to leave academia, he had already become a tenured professor of electrical engineering at the University of Pennsylvania.[18] He was recognized for his research on the ion-sensing phenomena in microfabricated sensors, charge transport in membranes, surface physics, and gas–solid chemical interactions.[19]

of the samples, and the need to enter and retrieve the results from the electronic medical record. Time is of the essence, as Lauks realized.

Hence, Lauks argued, it would be highly valuable to have a portable instrument that produces multiple measurements simultaneously in nearly real time. If he could realize that vision, then it would be possible to move away from centralized hospital testing facilities—which are sluggish and expensive—and enable much faster patient-centered testing by the bedside. This vision motivated him to leave academia, move to Canada, and found the start-up i-STAT. He was thirty-one. Based in Kanata, a suburb of Ottawa, in Eastern Canada, i-STAT sought to develop the first commercial microfabricated biosensor and the first general-purpose handheld blood analyzer. "There's a point in your life, when you're 25 to 35, when risk is exciting," Lauks is quoted saying of his decision to leave the university in 1983.[20] The initial system integrated an electrochemical chip (inside a cartridge) and a handheld reader with a pneumatic actuation system to displace liquids across the chip. This system performed six tests (calcium, sodium, potassium, dissolved CO_2 and O_2 concentrations, and hematocrit) in a small channel in ninety seconds—instead of fifteen to twenty minutes for larger analyzers. The user simply had to insert a cartridge into a handheld reader that electronically reported a user-friendly, quantitative readout of the results on the screen (figure 4.3).

The chemical sensor within the i-STAT used a detection principle based on ion-selective coatings that covered the electrodes. The engineers first used a photosensitive polymer called polyimide to insulate the electrodes along most of their length. They spin-coated the polyimide on top of the electrodes and patterned them with ultraviolet light to leave a small part exposed at the end of each electrode. The tip of each electrode was then coated with polymer thin films that had the property of being selectively permeable to different ions—a particular polymer thin film for each ion to be detected. To fabricate these ion-selective polymer coatings, they mixed a plastic with substances called ionophores that, in nature, carry specific ions across the membrane of cells. For example, to create a potassium-selective coating, they mixed plastic with valinomycin, a natural ionophore that carries potassium ions across the membrane of bacteria. They were able to create films that were selective for sodium, chloride, calcium, hydrogen, and ammonia ions, in addition to potassium. The i-STAT engineers deposited thin polymer films at the tip of the electrodes using a dispenser tool mounted on a precise positioner, very much like modern 3D printers. An array of these different electrodes on the silicon chip could detect various ions inside a small chamber that contained the chip. The chamber and microchannels were injection-molded in plastic.[21] Compared to the glucometer strips, which only measured glucose concentration, the i-STAT cartridge was much more demanding and complex to manufacture, but it also measured many more analytes.

The *New York Times* featured this widely successful platform in 1992: "Moving the Common Blood Test Closer to the Patient."[22] In 1998, Abbott Laboratories bought an 11.5 percent stake in i-STAT and implemented a five-year plan to distribute the technology, and in 2004 finally bought i-STAT for $392 million. By then, Lauks had already left (in 1999). In 2002, he started another company, Epocal (also in Ottawa), that focused on a similar line of tests (the epoc) but in a different format—inspired by plastic smart cards used in the banking industry—that was easier to manufacture. Having obtained US Food and Drug Administration (FDA) approval in 2006, the epoc was marketed as the cost-effective alternative to the i-STAT and was often able to replace it in hospitals, but its cartridges were more prone to failure. Lauks was right about the essence of time, but the per-assay cost and accuracy of the results provided by point-of-care assays (with higher reagent and cartridge costs) cannot compete with those provided by the high-precision equipment and lower reagent costs of centralized labs. Also, testing by clinical staff—who do not have laboratory work expertise—can compromise the reliability of the results.[23]

The i-STAT and the epoc have not replaced the central labs, but they are used ubiquitously in large hospitals to expedite urgent tests. Inverness/Alere acquired Epocal in 2013 for a reported $255 million, signaling the supremacy of the i-STAT. Previously, the use of the i-STAT handheld blood analyzer had broadened in 2008 when the FDA granted Abbott a Clinical

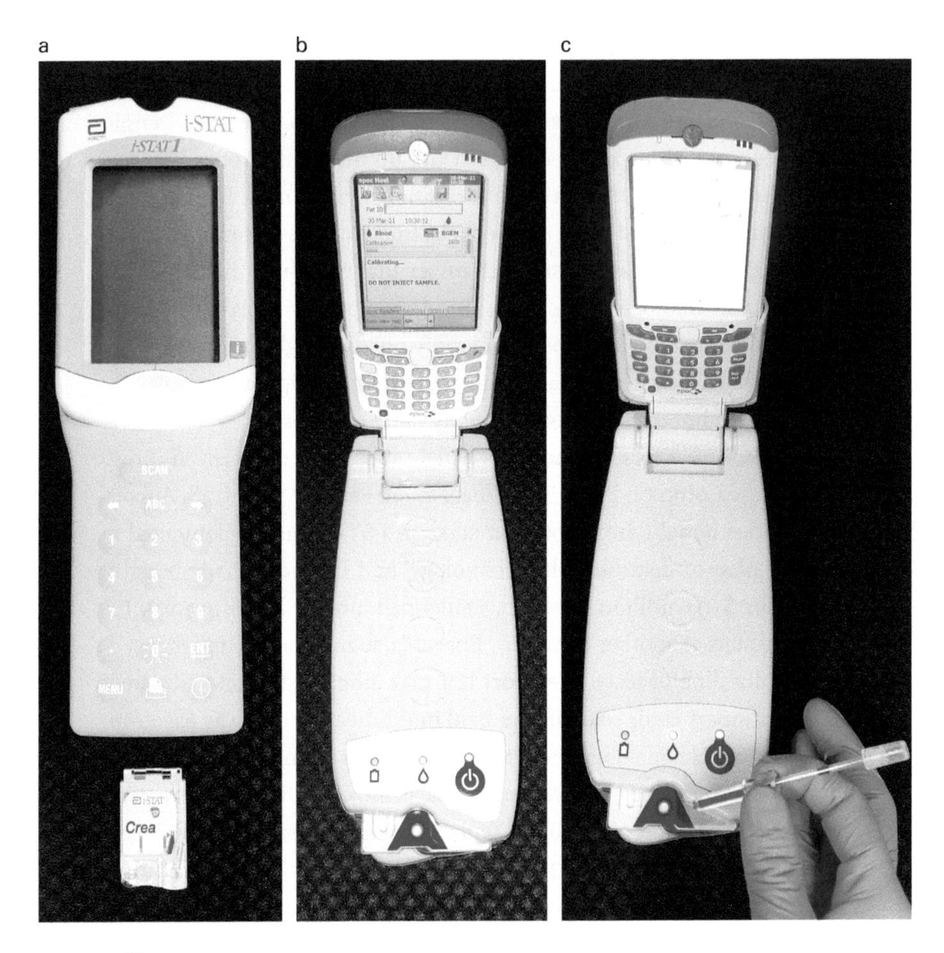

Figure 4.3

The first point-of-care microfluidic blood analyzers, both designed by Imants Lauks. (a) The i-STAT. A microfluidic cartridge for measuring blood creatinine (a by-product of muscle metabolism that is used as an indicator of kidney health) is shown next to the insertion slot. (b–d) The epoc, a wireless and more affordable version of the i-STAT. (b) An epoc shown with a microfluidic cartridge already inserted. (c) Loading of the epoc cartridge with test solution. (d) The cartridge after being removed from the epoc.

Images by Mark Wener, Department of Laboratory Medicine & Pathology, University of Washington.

d

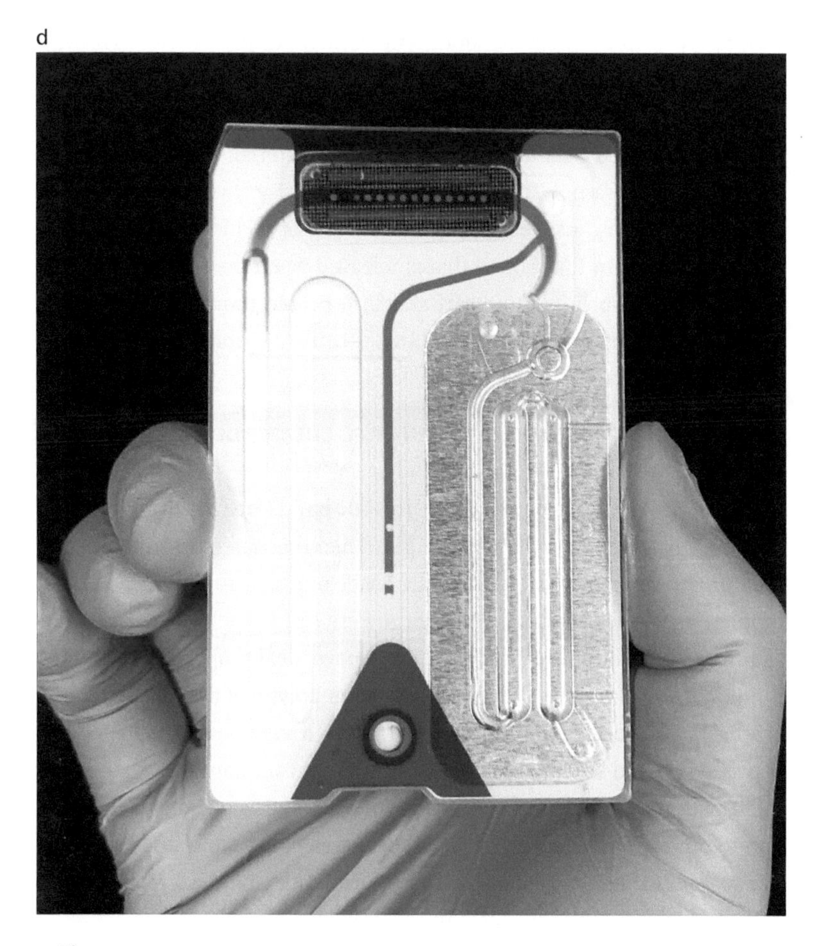

Figure 4.3
(continued)

Laboratory Improvement Amendment (CLIA) waiver for its CHEM8+ single-use diagnostic test cartridge. The CHEM8+ could simultaneously measure sodium, potassium, chloride, total carbon dioxide, glucose, urea, creatinine, ionized calcium, and hematocrit in arterial, venous, and capillary whole blood samples. The i-STAT now comprises twenty-five different tests performed in just ten minutes. These tests include antibody-based assays for troponin I (a protein that is released when the heart muscle has been damaged, such as in a heart attack) for rapid monitoring and diagnosing of a patient's metabolic condition—Lauks's vision from the start.

HOW THERMOPLASTICS ENABLED CHEAPER DEVICES

If you hold a FreeStyle or an i-STAT in your hand, the first thing you will notice is that they are not made of glass like the wafers used to decode the human genome. These devices are made of plastics—specifically, thermoplastics, a class of plastics that engineers can shape by melting them against a mold (box 4.7). Thermoplastics have ideal properties for microfluidics. Like all plastics, thermoplastics are polymers that chemists can synthesize in large quantities at a low cost—a kilogram of pellets costs between fifty and seventy-five cents. Companies can manufacture them in rolls or sheets. Microfluidic engineers can bond sheets together by lamination to produce multilayer devices. They are biocompatible, and many are transparent (unless a dye is added to color them). Chemical modification of their surface enables the attachment of biomolecules and cells.

There are several ways to shape thermoplastics with high precision. Lasers can be used to locally melt or ablate a thermoplastic using a digital design to produce prototypes at high speed (laser micromachining). An effective way is to mold them: once the mold has been fabricated, the cost per device is very cheap. One molding method consists of melting the plastic until it can flow and injecting it into the mold at high pressure to make sure the plastic fills up every corner of the mold (called injection molding; see figures 4.6, 4.7, and 4.8). The metallic molds for injection

BOX 4.7
The Pros and Cons of Manufacturing Thermoplastic Chips

When it comes to manufacturing, the low cost of the materials can be outweighed by the high cost of the equipment and processing. Plastic fabrication requires expensive instrumentation and expertise.

The most common approach to high-volume plastic manufacture, injection molding, is a process whereby molten plastic is injected into a metal mold using high pressures (figures 4.4 and 4.5). LEGO pieces are made by injection molding, for instance. It may cost tens or even hundreds of thousands of dollars to make the mold of an injection-molded device, but comparatively it takes only pennies for the materials to make each device from that mold. Therefore, plastic products are only worth manufacturing in large quantities when the return in sales justifies the capital investment. In practice, the corporate sector is risk averse: it only invests when it sees strong potential for profits or patents. This simple manufacturing principle applied well to the i-STAT or the FreeStyle because a large market guaranteed the sales of millions of devices. However, the same principle does not apply to microfluidic research in academic laboratories where designs are iterated by prototyping one device at a time. Hence, the vast majority of academic labs have adopted low-cost microfluidic prototyping techniques based on non-thermoplastic materials (see chapter 5). There is no doubt that thermoplastics have a vast potential for microfluidics, but, sadly, they have also caused a deep divide between industry and academics. Regarding microfluidic fabrication technology, industry and academics are on the opposite sides of a technological Grand Canyon.

Figure 4.4

In injection molding, the mold is held at high pressure between the metal plates of this tool.

Image courtesy of Holger Becker at microfluidic ChipShop GmbH.

(continued)

BOX 4.7 (continued)

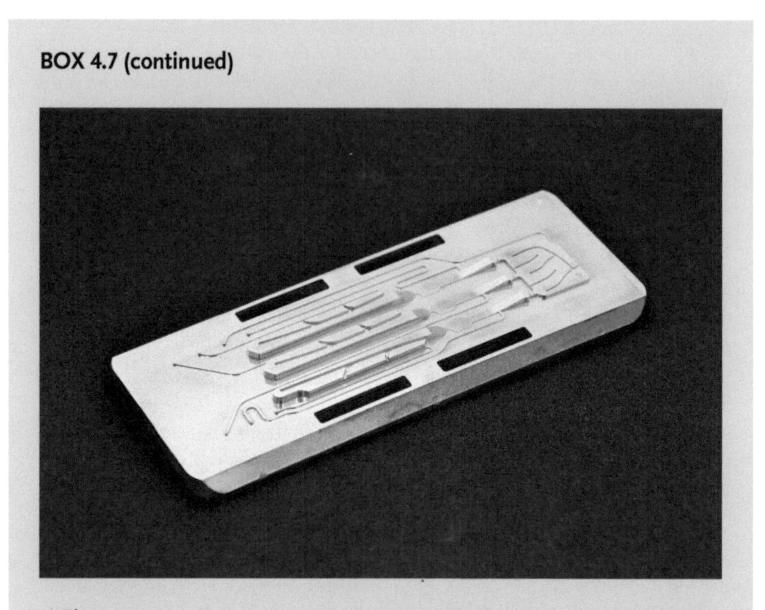

Figure 4.5
Because of the high pressures involved in injection molding, the mold itself needs to be made of a hard material such as metal. Since metals are difficult to pattern at high resolution, this requirement makes injection molding of microfluidic devices particularly expensive. This brass mold has been fabricated by micro-milling.
Image courtesy of Holger Becker at microfluidic ChipShop GmbH.

Most of the plastic objects that we use daily, such as pens and combs, are made of thermoplastics. You might wonder why it took so long for researchers to start making microfluidic devices with thermoplastics. Why didn't Steve Terry, Andreas Manz, and Lawrence Kuhn and Ernest Bassous make their devices directly in thermoplastics? The answer is easy: plastic microfabrication is not as easy as it seems—and it gets harder at smaller scales. Very few universities can afford to maintain a plastic fabrication service. Injection-molding machines are as big as a sofa, and people need a PhD in plastic manufacturing to operate them. Of the hundreds of injection-molding services in the world, only a few specialize in microfluidics and are concentrated in Europe (Germany, Switzerland) and Asia. If making large plastic parts is hard, imagine making parts that contain tiny details like microchannels and microchambers. A high-resolution mold costs many tens of thousands of dollars, so this is not a technique for prototyping.

Some low-cost alternatives to injection molding have emerged for plastic microfluidics fabrication. A desirable alternative is *dry-film resist lamination*, first proposed by Philippe Renaud in 2001. As opposed to traditional photolithography where the photoresist is applied in liquid form, a dry-film resist can be layered from a roll (so it can form the roof of a channel); after exposure, it is developed with a solvent developer and can be bonded by heat.[24] The combination of the plastic microchannels with metallization layers permits the fabrication of electrodes both inside and outside the channels. While this approach did not receive a lot of attention initially, it has grown in popularity lately. Alternatively, most groups have devoted their plastic prototyping efforts to more straightforward techniques such as molding of UV-curable plastics, computer numerical control (CNC) milling, laser micromachining,[25] or the pressing of a hot mold into a plastic surface (*hot embossing*). Starting in 2009, a combination of hot embossing and lamination approaches inspired a European international collaboration termed *Lab-on-a-Foil* that explored molding microfluidic chips on thin, flexible plastic sheet rolls similar to those used in food packaging.[26]

molding are, however, expensive. A low-cost molding technique involves pressing a hot mold into the plastic to leave an imprint of the mold features, a technique called hot embossing (also called thermoforming or imprinting). If there was one magical year for thermoplastic microfluidics, it must have been 1997. Within a short time, three papers reported these three different ways of fabricating plastic microfluidic devices: laser micromachining,[27] injection molding,[28] and hot embossing.[29] By that time, Imants Lauks had almost completed the development of the i-STAT and chose injection molding to manufacture the plastic microchannel around the sensor.

Injection Molding

Unfortunately, thermoplastic injection micromolding is usually out of reach of the budget of academic labs (box 4.7), where most innovations occur. A team from the company Soan Biosciences reported the first injection-molded microfluidic device in 1997.[30] They used microfabricated silicon masters to grow nickel microstructures electrochemically. They then used the resulting nickel replica molds to injection-mold

Figure 4.6
Various injection-molded microfluidic devices.
Image courtesy of Holger Becker at microfluidic ChipShop GmbH.

Figure 4.7
Injection-molded multistep microfluidic sandwich immunoassay device containing on-board pneumatic valves (not connected) and a vent membrane (the white area).
Image courtesy of Leanna Levine at ALine, Inc.

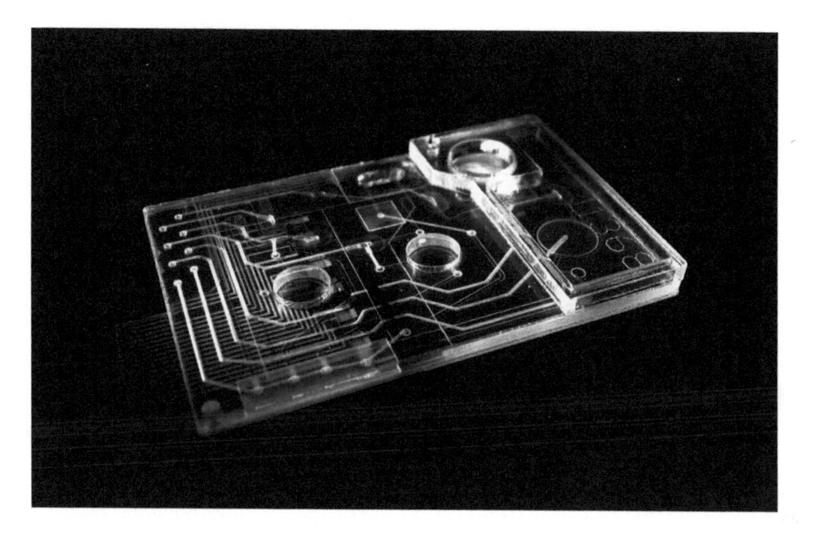

Figure 4.8
Injection-molded microfluidic immunoassay device for electrochemical detection. The two reservoirs at the center of the device are for sample introduction. *Image courtesy of Leanna Levine at ALine, Inc.*

plastic devices for high-resolution separations of double-stranded DNA fragments with total run times of less than three minutes. In 2005, Yoshinobu Baba's group performed genetic analysis on capillary array electrophoresis poly(methyl methacrylate) (PMMA) chips featuring ten 50-micron-wide, 50-micron-deep channels fabricated by injection molding.[31] Samuel Sia (box 4.8) and Vincent Linder, who met at Harvard working on the application of microfluidics to global health and diagnostics, have been thinking about injection-molded microfluidics for a long time.

Sia spent a month in Africa before joining George Whitesides's lab at Harvard to learn microfluidics and apply that knowledge to develop solutions for the Third World. Whitesides had been thinking along the same lines—a project he called Simple Solutions (chapter 7). There was one problem: Whitesides wanted Sia and Vincent Linder, another postdoc, to work on paper. "I thought the monolithic [non-paper] approach was more robust than paper, and to this day, I think there are limitations to paper,"

BOX 4.8: BIOGRAPHY
Sam Sia

Unlike most of us, Sam Sia's life has been influenced by global events and interests that shaped his career choices. Sam Sia was born in Hong Kong in 1975 when it was still under British control. In 1984, the UK and China signed the Sino-British Joint Declaration according to which Hong Kong would become sovereign after July 1, 1997. The Chinese government agreed to the one country, two systems principle and to keep Hong Kong's existing capitalist system and way of life unchanged until 2047. However, concern over these changes and the feeling that their kids would get a better education abroad led the Sia family to move to Canada in 1984 when Sam was eight years old.

During his undergraduate degree in biochemistry at the University of Alberta, Sam Sia did capillary electrophoresis research in the lab of Norm Dovichi, who had a large lab devoting much effort to the human genome. Next door was the lab of microfluidics pioneer Jed Harrison. Harrison had recently come back from Ciba-Geigy in Basel, Switzerland, where he had built the first glass microfluidic chips for capillary electrophoresis with Andreas Manz in 1992 (chapter 3).[32] In 1997, Sia began a PhD in biophysics at Harvard, working on peptide inhibitors that block HIV entry.

Just after his PhD in 2001, a month spent in West Africa defined how Sia would steer his career. "I wanted to think about how I could impact the world, and I wanted to learn about the world," says Sia. He went to Assahoun-Fiagbe, a remote village in southern Togo, two hours northwest of Lomé, the capital, to teach English to middle schoolers for four weeks. Togo is a tiny, sub-Saharan country of tropical climate that became independent of France in 1960, and has only enjoyed democratic governance, albeit with corruption, since 1993. The village, with 1,500 inhabitants, had no electricity or running water, so the living conditions were harsh. During his stay, a younger kid fell, and the wound became infected with tetanus, ultimately killing him. In the developed world where tetanus shots are ubiquitous, such a death would never have occurred. "It was a defining experience," remarks Sia.

says Sia. Linder agreed with him. Although they did some preliminary experiments in paper (chapter 7), they focused on immunoassay and global health experiments[33] that would be the foundational research for their start-up company. In 2004 they formed Claros Diagnostics, a global health and diagnostics company.

"We drove around Germany and Switzerland trying to find an injection-molding service that could produce our parts, but they were not able to," recalls Sia. Linder had to make many improvements—which he cannot fully disclose—to the injection-molding process that made the manufacturing faster and cheaper. The company has developed a microfluidic quantitative immunoassay finger-prick test for prostate cancer that is FDA-approved and meets the benchmark performance of much more expensive systems such as the Siemens Centaur. The microfluidic device (commercialized under the name Claros-1 Analyzer) is, of course, not made of paper—it is injection-molded. "If we start with something that is high performing, there are so many buttons we can push to make it cheaper," says Sia. "With the electronics and smartphone revolution, we don't want to throw that away." In 2011, Sam Sia's lab (using Claros's device) led a study to deploy HIV and syphilis immunodiagnostic microfluidic tests made of injection-molded thermoplastics in rural areas of Rwanda, entirely run without complex instruments.[34] The study was the first field trial of a microfluidic device in a developing country.

Hot Embossing

The easiest way to shape plastics is to start with a flat piece and press it into a hot mold. Although this technique, now known as hot embossing, has limitations (for example, through-holes are difficult), it is much simpler to implement than injection molding because it does not require the bulk of the plastic to flow—only locally soften at the surface.

Laurie Locascio (box 4.9) pioneered the hot embossing procedures that were rapidly adopted by many microfluidic engineers looking for low-cost alternatives to injection molding (box 4.10).[35] Locascio had read Jed Harrison and Manz's early papers and started working with silicon

BOX 4.9: BIOGRAPHY
Laurie Locascio

Laurie Locascio comes from a Sicilian-Irish family that settled in western Maryland. Her master's in biomedical engineering at the University of Utah addressed the area of chemically sensitive field-effect transistors and flow chemical sensors for implants. The National Bureau of Standards (now the National Institute of Standards and Technology [NIST]) offered her a position in 1986 to work with electrochemist Richard Durst. She would later manage to earn a PhD while raising three boys at home, but at NIST she was generously supported and promoted to lead a group of scientific researchers before obtaining her doctoral degree.

and glass. In 1995, she met with them at a conference and asked for advice because she could not get anodic bonding to work reliably. "That [anodic bonding] did not work for me—so we started working on plastics," she now happily recalls about the career choice she made. As they did not have access to an injection-molding system, Locascio's group first started experimenting with pressing hot wires into the plastic until their shape got imprinted or embossed—"that was easy"—and bonding the plastics—the challenging part. Together with Larissa Martynova, a guest researcher in her lab that year, they observed that "precise temperature control was critical to the success of the bonding," says Locascio. They published the results in 1997 in *Analytical Chemistry*.[36] Locascio also remembers how PMMA from a different lot suddenly produced inconsistent results, which was a source of frustration at the time. Yet, she adds with a smile as she tunes into her memories, the lack of batch-to-batch reliability "led to more research on how the surface chemistry of the plastic influenced electroosmotic mobility and protein adsorption." Electroosmotic flow was well understood in glass capillaries, where the glass surfaces were full of negative charges, but unreliable in plastic microchannels because the surface had an unknown composition. There was a lot to learn from plastics, and even from the frustrating imperfections of their manufacture.

BOX 4.10
Notable Hot-Embossed Chips

The vast majority of academic researchers—in part driven by the constraint of their need for low-cost prototyping—turned to non-thermoplastic molding techniques that were simpler, faster, and less expensive than silicon and glass micromachining (chapter 5). But a small number of academics deliberately pushed the challenges of working with thermoplastics. They focused their attention on plastics that were transparent, biocompatible, and easily moldable. Good examples are polycarbonate, PMMA (brand-named Plexiglas or Lucite), polystyrene, cyclic-olefin copolymer, and thermoplastic elastomer.

David Beebe argued for the fabrication of microfluidics in polystyrene for cell culture and drug testing because it generally features very low drug absorption.[37] Indeed, conventional cell culture plasticware is made of polystyrene.

Poly(methyl methacrylate) is a hard plastic with excellent biocompatibility, optical clarity from the visible to the UV wavelength, and low softening temperature (~100°C), while polycarbonate's durability and high softening temperature (~145°C) makes it ideal for DNA thermal cycling. In 2001, Steve Soper hot-embossed PMMA and polycarbonate plastic devices for DNA detection.[38] In 2010, Michael Murphy's group ingeniously designed a continuous-flow PCR device in a polycarbonate 96-well format.[39] Adam Woolley's group also used PMMA microfluidic devices to separate and extract multiple human serum proteins by electrophoresis.[40]

Advantageous properties of cyclic olefin copolymer (COC) include transparency, resistance to most solvents, low background fluorescence, and easy moldability to form microfluidic devices. As an example of a COC device, Lloyd Davis's team used COC microfluidic devices to detect bacteria by PCR in 2010.[41]

Finally, in 2011, Teodor Veres invented a thermoplastic elastomer (TPE), a plastic that can be hot embossed;[42] TPE is transparent (not to UV light) and biocompatible,[43] and has been used to fabricate devices with multilayer valves.[44]

As a technology, hot embossing is less expensive to set up than injection molding but results in larger cost per part so it is not appropriate for very-large-scale production runs. Various commercial foundries that specialize in hot embossing for microfluidics are available. These foundries typically serve startups that are not yet ready to pay for injection molding (>100,000 parts) but need from hundreds to 10,000 parts or more.

CAPILLARITY-POWERED MICROFLUIDICS

One striking feature about the FreeStyle and other glucometers is what they lack: pumps. How does the drop of blood get magically sucked in? If you go to any microfluidic laboratory, you will notice that most microfluidic devices must hook up to a rotary, injection, or gravity pump to power the flow. But we would not want to force diabetic patients to own and carry around a pump that runs the microfluidic device every time they need it. Indeed, that would be unnecessary: nature has provided a microscale pump on the surface of every fluid in the form of surface tension. The same surface tension of water that keeps the water strider afloat on a pond also causes water to spontaneously fill hollow tubes (even to ascend a capillary against gravity). We say that the fluid goes up "by capillarity" or "by capillary forces," although it is really the surface tension that pulls up the fluid (chapter 1). It is the surface tension at the surface of the drop of blood, which generates the force that draws that same drop of blood into the channel of the FreeStyle.

In the 1980s, Eastman Kodak Co. and other companies wrote a series of patents describing the first microfluidic devices that exploited capillary effects. Unfortunately, they did not publish the experimental data in peer-reviewed journals. The patents contain interesting descriptions of various micromilled capillary flow control elements, including microgrooves for promoting directional flow, abrupt geometry changes to stop the flow, sample dilution devices, and advanced capillary pumps with gaps that prevented liquid backflow.[45] As patents lack experimental detail and, back then, were not easily searched, academics largely ignored these developments.

David Juncker (box 4.11) and Emmanuel Delamarche are credited for first employing MEMS techniques to design microscale surface tension effects in 2002. Their autonomous microfluidic systems took fluid manipulation by capillarity a step further than simple capillary loading (as in the FreeStyle).[46] David Juncker's first project at IBM Zurich focused on capillary microfluidics (figures 4.9 and 4.10). These devices were made using

BOX 4.11: BIOGRAPHY
David Juncker

Juncker was born in 1973 in Aarau, a small town of a little more than 20,000 people in the foothills of the Jura mountains in northern Switzerland, half an hour west of Zurich. Another notable scientist, Albert Einstein, lived in Aarau for two years before entering the Zurich polytechnic school at seventeen years old. "Einstein was a bit of an inspiration because of his rebel side," says Juncker. He jokes that "I was the first in my class in primary school," quickly adding that his class in Aarau only had five kids. After Juncker studied electronics physics at the Institute of Microtechnology (IM) at the University of Neuchâtel (now part of the École Polytechnique Fédérale de Lausanne [EPFL]), he worked briefly in the biosensors lab of Elisabeth Verpoorte at the IM. Verpoorte, who had worked with Manz at Ciba-Geigy, was trying to figure out how electrically driven flow could generate pressure. After about a year and a half working and studying in Japan, the IBM Zurich Research Lab offered him a position in January 1999 to continue his PhD in the group of Bruno Michel. "At IBM, there were many senior staff you could learn from and a fantastic clean room with lots of support, which facilitated learning and helped me do all the microfabrication that we needed. It was a very supportive and collaborative atmosphere," he recalls.

poly(dimethylsiloxane) (PDMS), a flexible polymer that was becoming popular at the time among researchers (see chapter 5). The IBM group with Emmanuel Delamarche and Hans Biebuyck had published a paper in *Science* in 1997 demonstrating the first immunoassay using a PDMS microfluidic device;[47] the immunoassay involved capillary flow in 1-micron-wide microchannels. When Juncker joined the group, Hans Biebuyck had just left, but he built on Biebuyck's results. "I had worked with electrophoresis with 20-micron channels [in the Verpoorte lab], and it was very difficult to go smaller—when I saw the *Science* 1997 Biebuyck paper with 1-micron channels, I was very impressed!" says Juncker. Then Juncker came up with a way to hold the fluid inside the channel, a capillary retention valve, which was the critical element to making autonomous microfluidic capillary systems.[48] "I vividly remember my exhilaration at the time of the invention," he says proudly.

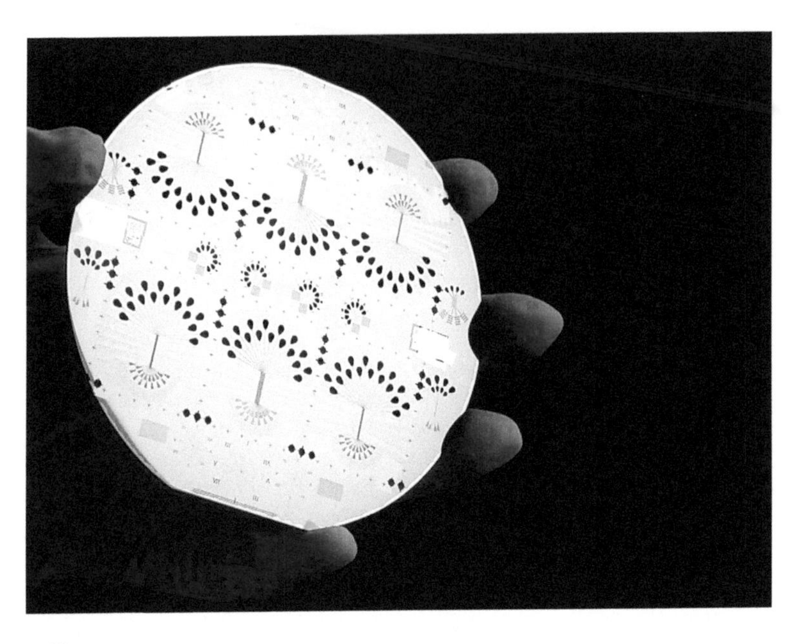

Figure 4.9
The photolithographic patterns in this wafer were used to replica-mold several capillary microfluidic devices in PDMS.
Image courtesy of David Juncker, McGill University.

The control and confinement of fluids by capillary effects became the core of Juncker's PhD thesis. Turning to look at the tall maple trees in the backyard of his house in Montreal, he adds: "Trees were an important inspiration—they pump water, and then they stop. Trees helped conceptualize the idea of autonomous capillary systems that both self-powered and self-regulated."

The capillary systems of Junker's thesis could only perform comparatively simple fluidic operations and were dependent on user intervention, and Juncker was resolved to fix that. Thin and sprightly, David Juncker points out with his charming Swiss French accent that "I don't like doing what everyone else does and, for me, science is about doing new things." In 2013, as a faculty member at McGill University in Canada,

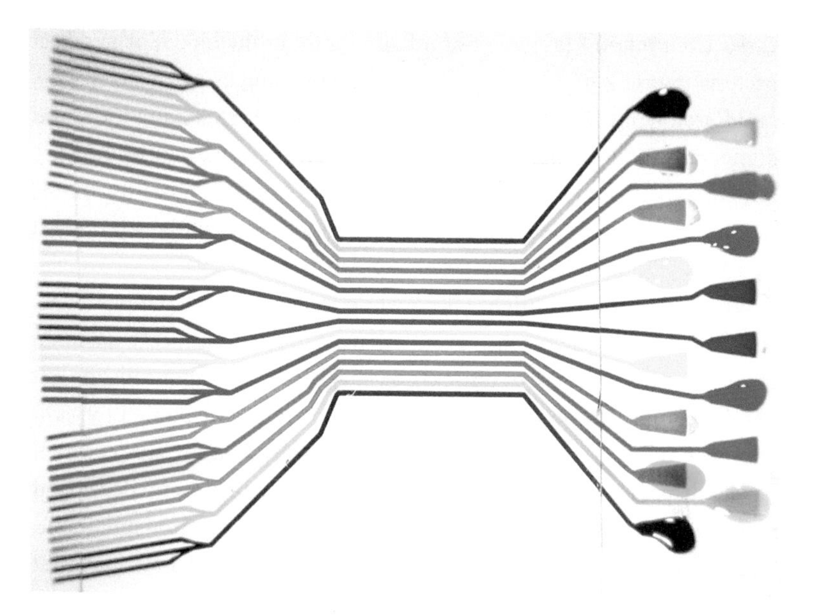

Figure 4.10
After drops of dye have been placed on the inlet pads on the right of this PDMS device, the dyes have filled the channels by capillarity—no pumps were required. *Image courtesy of David Juncker, McGill University.*

his group further elaborated the idea of autonomous capillary systems into what he calls "capillarics":[49] preprogrammed, self-powered microfluidic circuits that Juncker's group builds in modules from fundamental microfluidic elements, much like electrical engineers build more complex circuits from simple elements like transistors, capacitors, and resistors. His capillary fluidic elements include capillary pumps, trigger valves, retention valves, retention burst valves, flow resistors, inlets, and vents. "With capillarics, we can aspire to make devices that can perform really complex operations," says Juncker. As a demonstration, they built a circuit that, following the sample loading by the user, autonomously delivered a preprogrammed sequence of multiple chemicals according to a predetermined flow rate and time to measure the concentration of

C-reactive protein[50] (a marker of inflammation in the blood) or to detect bacteria (figure 4.11).[51] Juncker now sees 3D printing as a transformative, rapid-prototyping technology that will enable even higher complexity and democratization of capillarics.[52] "With 3D printing, we can embed the structures and the logic in a 3D structure and integrate more functions, we don't need a clean room, and working designs can easily be reproduced by others based on the 3D models," explains Juncker.

Whereas capillarics might take a while to develop into the platforms envisioned by Juncker, other groups have exploited surface tension in other ways that are more intuitive than capillarics. Applications that do not require complex programming could benefit from more straightforward approaches such as hanging microdroplet arrays (figure 4.12), which can be dispensed by robotic pipettors. Shu Takayama's group developed injection-molded 384-well plates that allowed for the creation and maintenance of miniature cell cultures in the form of upside-down hanging

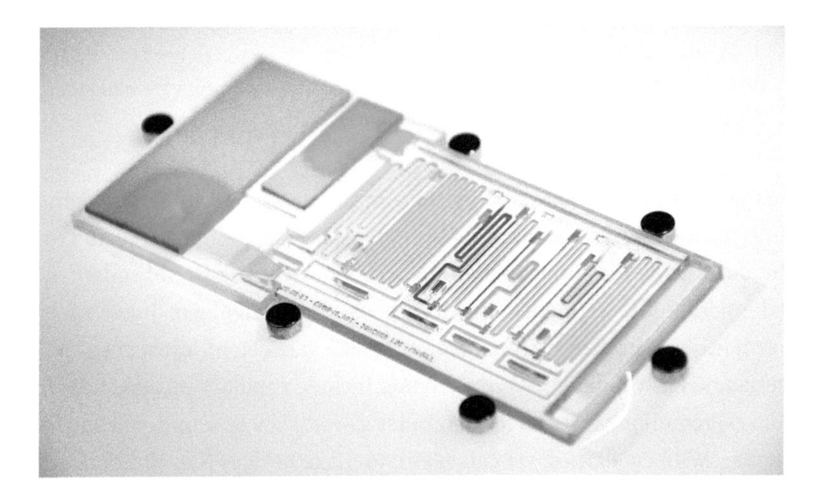

Figure 4.11
"Capillarics": a self-powered microfluidic device containing capillary valves capable of performing an assay based on a sequence of preprogrammed operations.
Image courtesy of David Juncker, McGill University.

Figure 4.12
Array of hanging microdroplets.
Image courtesy of CBMS (Okaar Photography).

drops. Not finding a solid surface, the cells in the droplet aggregated together to form a three-dimensional "organoid" that was physiologically more similar to normal tissue than traditional 2D cultures attached to plastic.[53] Hanging drop systems are usually targeted for large pharma laboratories and require expensive robotic dispensers. While these pipettors are expensive, they are easy to program and operate.

Others, like David Beebe's group, utilized the surface energy present in a small drop of liquid to pump the liquid through a microchannel.[54] By carefully balancing the sizes of the drops dispensed at the inlet and the outlet, it is possible to pump fluids in either direction. Counterintuitively, smaller drops pump more strongly than larger drops because their smaller radius of curvature makes their surface tension higher.[55] Based on this concept, Beebe published high-density "tubeless" microfluidic device arrays for automated cell culture, but they are rather static devices.[56] In contrast, Juncker's capillarics run pre-programmed sequences with a simple contact with the fluid—they are stand-alone like the trees that inspired them.

THE LABCD

Microfluidic engineers that develop point-of-care instruments look at the smartphone with envy and wish they could have something as portable and small as the battery to pump fluids around. Surface tension and capillarity do fulfill that role, but the energy stored on the surface of water, released once water comes in contact with a suitable interface, cannot be used in an already-wetted channel. An alternative strategy is to drive the flow by *centrifugal force*, a notion that has independently occurred to people more than once.

The beginnings of the idea to exploit centrifugal force can be traced back to medical devices that predate microfluidics. In 1972, a team from Oak Ridge National Labs (ORNL) demonstrated a miniature fast analyzer, consisting of seventeen cuvettes attached to a plastic rotor that rotated at speeds up to 5,000 revolutions per minute. They used the centrifugal force to transfer and mix a series of samples and reagents into the cuvettes; an optical detector monitored the reactions in real time.[57] Abaxis bought the ORNL patent and, in 1995, announced the Piccolo, a clinical blood analyzer based on a rotating PMMA disk with capillary channels. For each sample, the Piccolo processed 90 microliters of whole blood, separated the red blood cells by centrifugation, generated multiple aliquots of diluted plasma, and reported a dozen tests in fourteen minutes.[58]

Unaware of the Piccolo, a Harvard biologist working on DNA technology, Steve Higgins, also came up with the idea of implementing centrifugal force to power a microfluidic chip in October of 1995, the same year the Piccolo was announced.

Great lessons in entrepreneurialism and perseverance can be drawn from the decade-long adventure to commercialize the idea, undertaken by Higgins and his lab mate, Alec Mian (box 4.12). After Mian completed his PhD, he started Gamera Bioscience in 1992 and invited his Higgins to join as "the technology guy." They rented lab space for the new company by the Charles River. They needed a physicist, so Mian's girlfriend introduced them to her friend's boyfriend, Gregory Kellogg, a

BOX 4.12: BIOGRAPHY
Steve Higgins and Alec Mian

The son of a career naval officer, Steve Higgins was born in 1957 in Coronado, California, where his father was stationed. After his father's retirement, the family settled in Maine. Upon graduation from high school, he also joined the Navy "and used the GI bill to get my degree," says Higgins referring to the program that pays for the college expenses of US soldiers. In 1987, after an undergraduate degree in biology and a master's in biotechnology, both at University of Massachusetts Boston, he became the first hire of George Church, who was then setting up his lab in the Department of Genetics at Harvard Medical School. Church wanted to develop a high-throughput genome sequencing method called multiplex DNA sequencing. He hired Higgins as a technician, but "George treated me like a graduate student," Higgins recalls. Higgins wrote all the computer program for a DNA synthesizer. In 1988, Church and Higgins published a paper on multiplex DNA sequencing in *Science* and shared the patent on it.[59] They used the method to sequence the genome of the first organism (*H. Pylori*).[60] In the Church lab, Higgins met Alec Mian.

Alec Mian grew up in Montreal, Canada, with a Pakistani father and a German mother. He studied neurobiology at McGill University, Montreal. He started a PhD in genetics at King's College at the University of Cambridge, UK, but his girlfriend at the time was an elite athlete training on the US National Rowing Team out of Boston; therefore, he transferred his PhD to George Church's lab at Harvard in 1989. Mian was very entrepreneurial, so he started a company in Boston, Gamera Bioscience, after completing his PhD in 1992 to develop gene amplification technologies.

PhD in physics from Harvard who was doing a postdoc in materials science at Massachusetts Institute of Technology (MIT); they hired Kellogg. Although the company's initial focus was gene amplification, they identified microfluidics and laminar flow as a technology enabler for DNA processing. In October 1995, it suddenly occurred to Higgins that "if we could spin and control the rpm [revolutions per minute], we could do a lot of things with centrifugal force." Alec Mian suggested the round, CD format (compact disks used for digital audio) because it could be rotated, micromolded with microchannels, mass produced, and combined with a laser to interpret results. "The CD features would broaden the scope of the device under development to process not only DNA samples but also biological samples and other supporting lab assays," says Mian. Because it looked like a CD, they called it *LabCD* (figure 4.13).

But the vision was too far ahead of the technology. Alec Mian realized he needed to recruit the help of microfabrication experts. In 1997, Marc Madou, then a professor at Ohio State University, published his textbook *Fundamentals of Microfabrication*, which became an instant bestseller in the MEMS community, so Alec Mian asked him if he would consult for Gamera. Madou agreed. Years later, this consulting role would become crucially important for the development of the LabCD concept.

Marc Madou and Greg Kellogg published the first paper on the platform in 1998. As a plastic disc containing microchannels rotated, the centrifugal force and inertia of the fluid provided the pumping force.[61] They also realized that capillary burst valves—an open-ended channel that pins the flow by surface tension and, with an extra push of pressure, the flow can be caused to burst through—could be selectively activated at different speeds by changing the location and design of the valves. The disc-based blood-plasma separation reported in the paper was a significant achievement. The same principle could be extended to fluids with a wide disparity of viscosities and properties (from blood to milk and urine), and it was not surface-sensitive like Manz's electroosmotic separation. Yet it was initially met with skepticism. Intuitively, it seemed too complicated to rotate a whole "lab." For starters, it was difficult to watch the channels while they were rotating—one had to use a stroboscopic camera to view what was going on inside the chambers.

Most tragically for Gamera, the company could not secure a foundational patent on the concept of centrifugal microfluidics because centrifugal analyzers such as the Piccolo had existed for a while, and "the concept of using centrifugal forces for moving liquids around was well established," concedes Higgins. Mian started looking for ways to manufacture the platform and approached Amersham Pharmacia Biotech in 1997, who was making an injection-molded capillary electrophoresis device using the

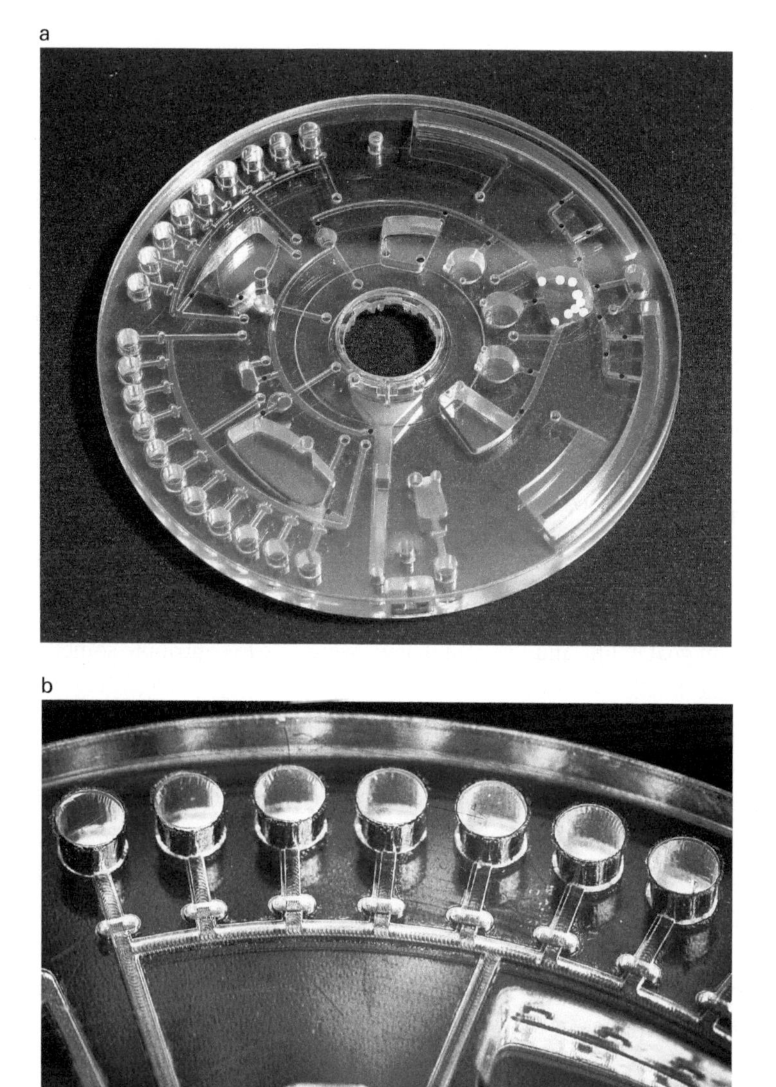

Figure 4.13a

A 12-centimeter-diameter LabCD designed by Christopher Ko's lab that has been injection-molded in cyclic olefin copolymer. The device, which performs immunoassays from whole blood, features microvalves that contain a ferrowax substance that can be melted by a laser to open or close the valves.

Image courtesy of CBMS (Okaar Photography).

Figure 4.13b

Detail of figure 4.13(a). The widenings in the radially oriented channels act as "burst valves." When flow arrives at the widening, it becomes pinched by the walls, requiring additional centrifugal force to restart the flow.

Image courtesy of CBMS (Okaar Photography).

CD footprint. Realizing that Gamera did not have a foundational patent, Pharmacia instead started a spin-off on their own called Gyros in 2000. "One of our problems was finding the right application, and Gyros has done an amazing job at developing immunoassays. It is perfect for the platform and something we should have done. I salute them," says Higgins now.

That he does not hold any grudges against Gyros speaks very highly of Higgins. Perhaps he got the better deal—history may remember him as one of the coinventors of centrifugal microfluidics (the format is different from the Piccolo's), and at Gamera, he met his future wife. In his backyard full of tall, lush green trees in Maryland, after sipping a glass of white wine, he muses: "We had an idea that changed the way people think about microfluidics. Making a whole lot of money was never my main goal in life, and I'm glad that we were able to contribute something to the world."

Although Tecan acquired Gamera's system in 2000 and—as if to prove the skepticism right—discontinued it in 2005, Madou (now at University of California, Irvine) persisted in its development.[62] "My primary motivation was to get my students a good PhD thesis," humbly points out Madou when asked how he has benefited from his work. As the user base and applications have expanded, the initial vision of Mian and Higgins has also grown (box 4.13). "I'm now looking at it as a hospital-replacement instrument that does many of the tasks that are often relegated to separate expensive pieces of equipment," explains Madou.

ELECTRONIC CONTROL OF FLUIDS

Engineers have long envisioned the use of electronics to control fluids. Electronics have several convenient features: they are readily programmable, inexpensive, and widely available. The main challenge has been to make water obey electrical commands. After all, pure water is an insulator. Fortunately, water molecules have charges that can interact with charges present on a surface. Engineers can place an electrode under a

BOX 4.13
The LabCD, a Mature Platform

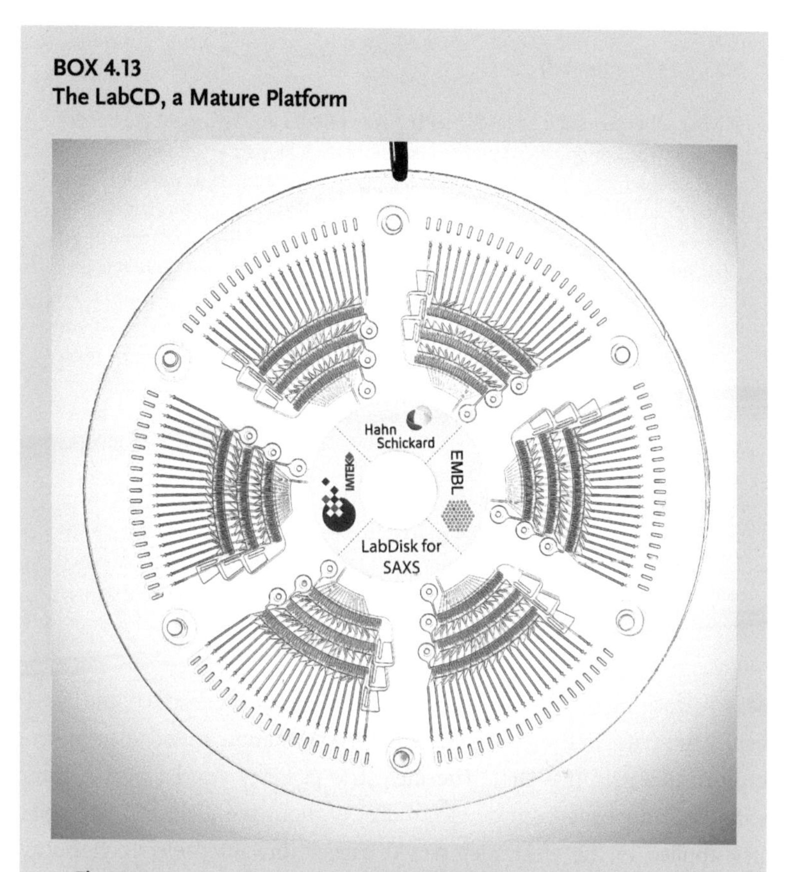

Figure 4.14
A centrifugal microfluidic LabCD platform for protein structure analysis using high-energy X-rays. One LabCD prepares 120 different measurement conditions, grouped into six dilutions. Each of the six segments includes the three input liquids, the combination, and the mixing in different predefined concentrations. The mixtures then reside in the readout chambers. Readout can be performed on disk in the X-ray facility. *Image courtesy of Roland Zengerle and Daniel Mark at Hahn-Schickard and the University of Freiburg.*

By the time the microfluidics community realized that other valved systems also had unfriendly requirements—such as many inputs and poor manufacturability—the LabCD had become a mature, push-button diagnostics platform with many users and developers around the world (figure 4.14).[63]

(continued)

BOX 4.13 (continued)

Valving schemes, such as laser-melting wax valves[64] and hydrophobic-barrier valves,[65] were developed. Marc Madou's PhD student Nahra Noroozi created a mixer design in 2009 that ended in an empty chamber so that, as the inlet fluids became mixed in an intermediate chamber by the centrifugal force, they also compressed the air in the empty chamber. As soon as she stopped the centrifugal force, the fluids were released back by decompression; the repetition of this procedure caused the rapid mixing of the inlet fluids.[66] Dario Mager's group from the Karlsruhe Institute of Technology has controlled electrochemical reactions within the LabCD platform by wirelessly coupling electrical power into the rotating disk.[67] Roland Zengerle and Jens Ducrée from Freiburg's Institute of Microsystems Technology (IMTEK) in Germany have used cyclic olefin copolymer disks to develop LabCD blood analyzers with colorimetric output.[68] A Samsung team has demonstrated an injection-molded PMMA LabCD that detects hepatitis from whole blood by performing an immunoassay on the disk (figure 4.13).[69] The list is very long. The LabCD platform has also been deployed in low-resource settings[70] and inspired inexpensive, human-powered devices such as a paper-based centrifuge[71] or the fidget-based centrifuges for blood[72] and urine[73] analysis.

glass surface to change the amount of charge on the surface and, as a result, tune its wetting properties—a phenomenon called electrowetting on dielectric (EWOD). The dielectric (a fancy word for insulator) is the layer deposited on the electrodes. Without the dielectric, the voltages applied to the electrodes would trigger unwanted electrochemical reactions in the fluid. Microfluidic engineers have designed a variety of devices to exploit the phenomenon of EWOD. They all share an architecture of parallel glass plates separated by a thin space within which a grid of addressable microelectrodes shuffles droplets around.

EWOD is both simple and incredibly powerful. It harnesses the programmability and electrical addressability of microelectronics to control the motion of discrete fluid volumes without mechanical parts. EWOD devices typically consist of multiple discrete regions (figures 4.15 and 4.16). Using a software interface, users can program how the droplets (loaded from special ports) move from one region to another and how they want

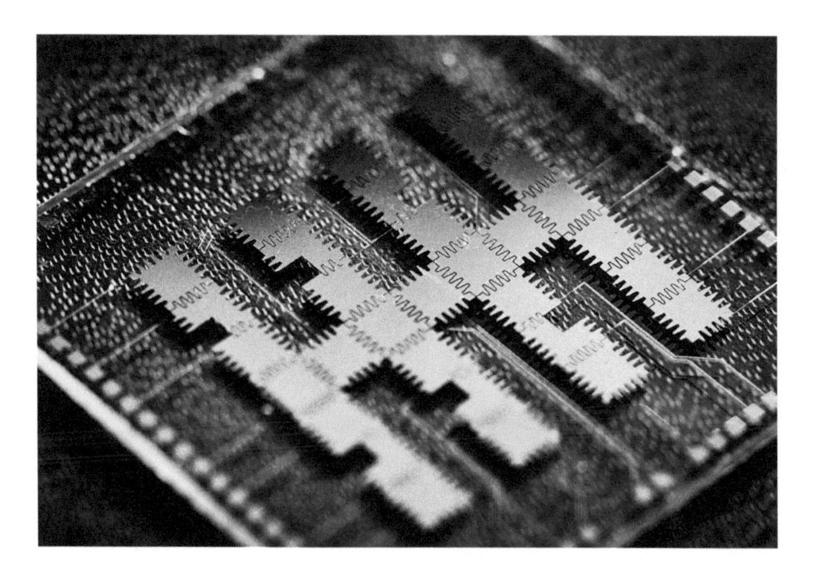

Figure 4.15
EWOD device showing an array of addressable electrodes.
Image courtesy of CBMS (Okaar Photography).

to use the different regions for any given assay, such as a fluorescent readout (figure 4.17) or a microscope-based assay (figure 4.18). Add wireless control and you have the dream of just about every microfluidic engineer.

As happens to many good ideas, multiple groups created EWOD microfluidic devices in one incarnation or another. In 1983, J. Samuel Batchelder designed a dielectrophoretic manipulator that consisted of two glass plates featuring opposing transparent planar electrodes made of indium tin oxide. The device could manipulate discrete, 1-millimeter-diameter water droplets in oil, mercury droplets in air, and air bubbles in hexane, but more than 250 volts were necessary.[74] Others reported an electrowetting micropump[75] and the actuation of water droplets in open air.[76] Between 2000 and 2002, teams led by Richard Fair[77] and Chang-Jin (CJ) Kim[78] raced to perform EWOD with air-surrounded droplets between two glass plates. Fair's group, whose platform was commercialized by

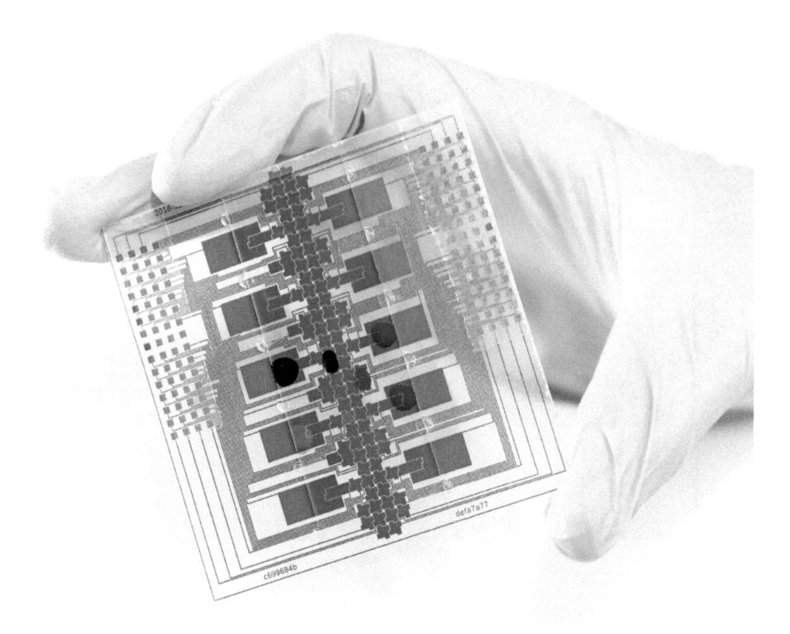

Figure 4.16
EWOD assays are portable and user-friendly. The image shows a device with four loaded color-dyed samples.
Image courtesy of Aaron Wheeler, University of Toronto.

Advanced Liquid Logic (acquired by Illumina), recently demonstrated DNA sequencing using a more advanced version of the platform.[79] Aaron Wheeler's group and colleagues championed CJ Kim's system to produce programmable immunoassays,[80] DNA amplification and analysis,[81] cell-based assays,[82] and cell cultures,[83] among others,[84] a platform now being sold by Miroculus. Recently, Emmanuel Delamarche's group has used dry-film resist lamination to develop a miniature, Bluetooth- and smartphone-controlled version of EWOD chips. With a few volts, Delamarche's chips can run several assays by capillarity from a single drop deposited at the inlet[85]—perhaps the closest anyone has gotten to achieving the vision of a "fluidic smartphone" for point-of-care assays.

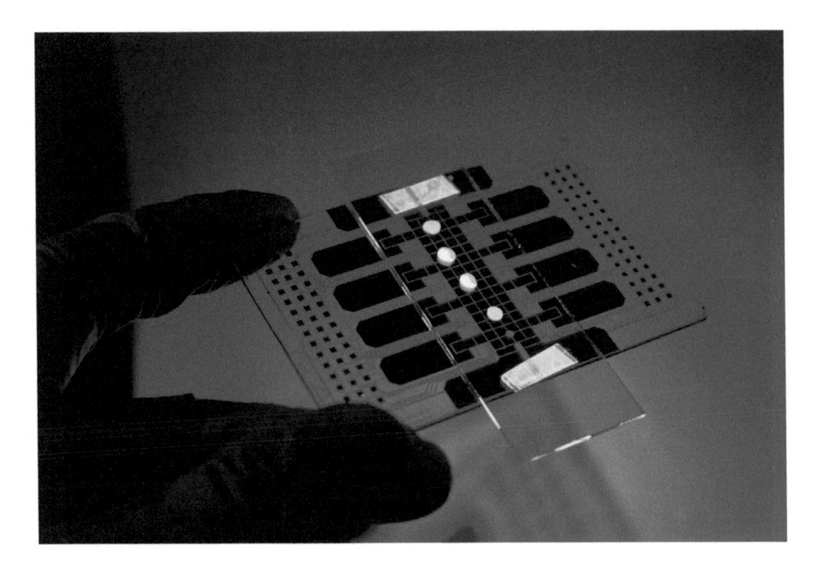

Figure 4.17
EWOD devices are compatible with fluorescent samples.
Image courtesy of Aaron Wheeler, University of Toronto.

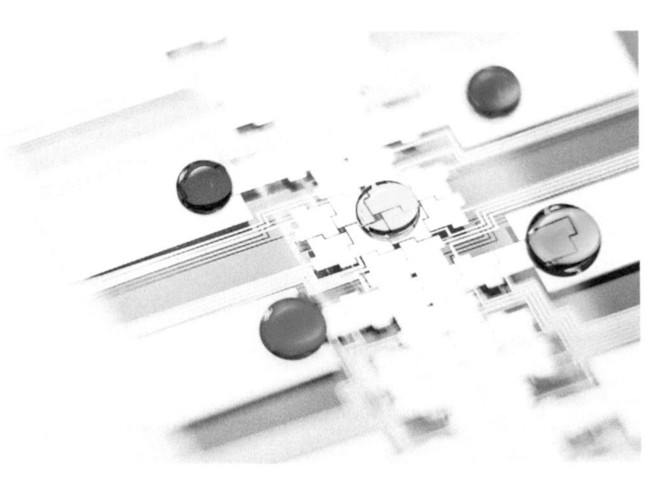

Figure 4.18
EWOD devices can easily be inspected in real time with a microscope.
Image courtesy of Aaron Wheeler, University of Toronto.

WEARABLES

The materials challenges and manufacturing constraints posed by point-of-care devices have been an enduring lesson for the field at large—one that has influenced the fabrication of more complex devices, such as organ-on-a-chip (chapter 6) and even implantable microfluidic devices.[86] But have you ever wondered whether it would be possible to design a thin patch that analyzes tiny drops of your sweat as you go for a run to monitor your fitness? The short answer is yes—mostly thanks to the pioneering work and immense creativity of John Rogers (box 4.14).

Rogers's colossal success has hinged on exploiting a rather ordinary and well known microscale observation: a microelectronic chip's active part (where electrons are flowing) represents an exceedingly thin layer confined to the electronic chip's surface—only a fraction of a micrometer

BOX 4.14: BIOGRAPHY
John Rogers

Rogers's father has a PhD in atomic and molecular physics and his mother, Pattiann Rogers, "is a widely published poet with an emphasis on physics and the natural world," as Rogers mentions proudly. Born in 1967 in Rolla, Missouri, Rogers still looks like an unassuming, shy, small-town college graduate. But the man underneath follows on his family achievements. He double majored in physics and chemistry at the University of Texas at Austin, did a PhD at MIT in physical chemistry and a postdoc at Harvard University, and took positions at Bell Labs and the University of Illinois before moving to Northwestern University in Chicago. There he leads the largest materials science academic laboratory in the United States, with more than one hundred scientists at any given time. In 2019 alone, his group published around seventy papers and filed seventeen patent applications. Most groups throw a big party when they land a paper in a top journal like *Nature*, *Science*, or *Cell* because it usually means not only recognition but also follow-up funding. His group averages dozens of high-profile publications every year and an average of one journal cover *each month*. The list of Rogers's awards is several pages long and departments around the world fight to hire his trainees as their faculty. As a testimony of his legacy, eighty-three of his alumni had become professors as of 2020.

thick. The inactive silicon part of the chip below the active surface, which is about a thousand times thicker than the active part, is as electronically dead and silent as a rock. Therefore, as John Rogers argued more than fifteen years ago, one might as well get rid of the inactive part and grow or mount the electronics on flexible polymer films to enjoy the benefits of plastics (figure 4.19). Plastics are transparent and biocompatible, and can be bent or inflated into various shapes. Rogers modestly describes this so-clever, multimillion-dollar idea as a "natural" one, although it would be hard to find another materials scientist alive who has had a bigger impact than him. "We were previously working on thin polymer semiconductors on plastic sheets as the basis for flexible circuits—so maybe it was natural to think about the active stuff on the top surface and the mechanical support below," Rogers explains. "We just needed to replace the wafer as a mechanical support with our plastic sheets, retaining the thin silicon near the surface," he adds.

Rogers became interested in flexible electronics during his stint at Bell Labs but became aware of the biomedical applications later. "We

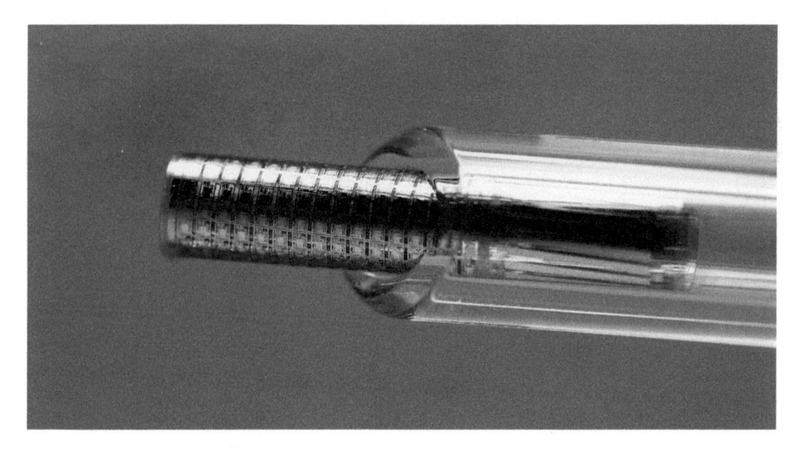

Figure 4.19
Flexible electronics fabricated on biocompatible plastic substrates and mounted on medical devices have been used to detect bioelectric activity from neural and cardiac tissue in vivo.
Image courtesy of John Rogers, Northwestern University.

were working on technologies for flexible displays in those days," he has said.[87] "Electronic newspapers, something like that . . . That's kind of how I got my start in flexible electronics. When I moved from Bell we were approached by a group of neuroscientists at the University of Pennsylvania wanting to know if we could take our flexible display circuits and put them on a brain." These thin films can be bent, folded, buckled, or—with the right patterns of loops and bends—stretched into functional constructs like micro-origamis,[88] and thus built into novel 3D devices like microelectronic circuits, biomedical sensors, antennas, and microrobots, among others. Taking advantage of materials buckling, Rogers's group unfolded patterns into fanciful nanoflowers and picnic microtables[89] (figure 4.20) that we thought were impossible to fabricate a few years ago. Using this simple strategy, they mounted thin-film electronics on tattoos to detect bioelectric activity wirelessly ("epidermal

Figure 4.20
This metallic picnic microtable is just 1 micrometer wide. The table is originally patterned flat but pops up when the substrate is compressed from all sides. *Image courtesy of John Rogers, Northwestern University.*

electronics"[90]), an idea they have extended to the wireless monitoring of neonates in intensive care.[91] They also performed electrocardiograms by placing a sensing array of electrodes on the surface of a beating heart[92] (instead of on the skin over the chest). Dissolvable silk substrates allowed electrodes to conform to the cavities of the brain surface and obtain recordings from an epilectic patient.[93] Balloon catheters with incorporated stretchable electronics enabled the electrical monitoring of catheter-based heart operations (figure 4.21).[94] These are just a few science-fiction-like examples that have expanded the imagination of the

Figure 4.21
Balloon catheters are thin tubes that are introduced into the body, such as into a blood vessel to release a blood clot, upon inflating the balloon. The presence of flexible microelectrodes on the balloon allows for monitoring the procedure in real time electronically.
Image courtesy of John Rogers, Northwestern University.

community beyond the imaginable. "To be able to put electronics on the human body occurred to us as being a richer area for research and a lot more promising for broad societal impact than a new piece of consumer electronic gadgetry," Rogers said.[95]

Rogers has extended this idea to wearables that incorporate thin (less than 1 millimeter thick) microfluidics for sampling sweat. For the same price, microelectronics technology (see box 1.3 in chapter 1) allow for integrating thin-film sensors that measure temperature and heart rate. In 2016, his group demonstrated a temporary tattoo whose thin PDMS microfluidic channels, sealed onto the surface of the skin, could harvest sweat from pores on the epidermis.[96] Using a combination of capillarity and the natural pressure (~70 kilopascals) generated by perspiration, the device then routes the sweat to various reservoirs for colorimetric sensing of chloride and hydronium ions, glucose, and lactate. Electrochemical sensing is also possible.[97] The battery-free electronics use near-field communications (NFC, the short-range equivalent of Bluetooth that allows you to "tap" payments with a cell phone) to harvest power wirelessly and

Figure 4.22
Microfluidic wearable for chrono-sampling and analyzing of sweat.
Image courtesy of John Rogers, Northwestern University.

to relay quantitative values of sweat rate, total sweat loss, pH, and concentrations of chloride and lactate to a smartphone nearby[98]—all in a thin tattoo. A similar sweat patch that incorporated capillary burst valves allowed twelve timed sweat samples (one per minute, see figure 4.22).[99] They cleverly organized the twelve positions of the sensor in a watch-like manner. Nothing but NFC and capillarity forces power the device, thin as your skin.

* * *

Thanks to these engineers, we will all be able to sport cyborg microfluidics on our bodies very soon to monitor our physical exercise routines. Still, when John Rogers explains to me that "our sweat microfluidic systems are being launched commercially through Gatorade, at Dick's sporting goods starting early 2021," I need to pinch myself for fear of living in the sci-fi movie *The Matrix*. If you like the idea of sticking electronics on your body, and you are already annoyed that the sensors will not function unless they have an NFC source nearby, Wei Gao's group at Caltech has recently developed a solution to this problem: they have developed tattoos with lactate biofuel cells that generate power for the chip from the perspiration itself.[100]

Diabetic patients like my friend Jeff, meanwhile, might have even wilder hopes. It should be possible, in principle, to place a microneedle on a flexible chip that samples their blood to monitor their glucose levels continuously and wirelessly transfer the results to their smartphone. An array of chemical microsensors could be integrated for no extra cost to measure various blood parameters just as the i-STAT does. But anybody can dream of a better future—to actually get there, we need microfluidic engineers like John Rogers. A large team led by him and Tyler Ray (University of Hawai'i at Mānoa) has recently developed "sweat stickers" to measure sweat chloride from children as a way to diagnose cystic fibrosis, a common life-shortening genetic disorder, in a child-friendly way.[101] Our kids' generation compared their smartphones at the school playground; their kids might soon be trying to match each others' microfluidic wearable patches.

5 THE DEMOCRATIZATION OF MICROFLUIDICS

At the birth of a technology, its access is often circumscribed to a small group of individuals. Borrowing political language, one could say that an aristocratic elite controls access of the users to the technology at this first stage, and it is not until later that it becomes democratized. The car, the telephone, and the computer, were once restricted to a high society, but the Ford Model T, the PC, and smartphones democratized transportation, net access, and mobile telephony, respectively.

Microfluidics followed a similar pattern. At the beginning of the 1990s, microfluidic engineers were a tiny community that had restricted access to a unique set of tools. Most of the engineers—those that had the privilege of access to a clean room—made their devices in silicon or glass using photolithography and etching. We have seen examples of the pioneering work by Lawrence Kuhn and Ernest Bassous in the 1970s on silicon-etched inkjet nozzles (chapter 2) and by Andreas Manz in the 1980s on glass-based capillary electrophoresis devices (chapter 3). Imant Lauks was one of the few who had started thinking about manufacturing microfluidic devices at high throughput with plastics by injection molding in the 1980s (chapter 4). But all these fabrication techniques required a prohibitively expensive setup, which meant that most labs—having no access to a centralized facility—could not make use of microfluidic technology.

In microfluidics, what caused the aristocratic privileges to end was the new use of a polymeric material called PDMS, in particular PDMS molding for the fabrication of devices. The story of this chapter is not about a particular application; rather, it tells how PDMS molding, and subsequently 3D printing, not only radically changed the devices but also caused a blossoming of new activities by the many scientists that were itching to use them.

PDMS: A DEMOCRATIZING MATERIAL

Poly(dimethylsiloxane) (*PDMS*), a polymer that looks like transparent rubber, is a clear form of the silicone that you may be familiar with in toys, kitchen utensils, and bathroom appliances. It is elastic, and there is no smell to it. Very biocompatible, PDMS is commonly used in medical devices and breast implants. The polymer is formed of many units of a viscous liquid monomer (dimethylsiloxane) that is used in a variety of products that you might be wearing right now or touched today: cosmetics, shampoos, conditioners—which give hair that shiny and slippery feel—lubricants, adhesives, caulk, heat transfer fluids, aquarium sealant, silly putty, and mold release agents. You might even be surprised, or disgusted, to learn that many fast-food businesses utilize dimethylsiloxane (at just ten parts per million) in fryer oil as an anti-foaming agent.

PDMS-based fabrication is based on molding. Fabrication of the mold for microfluidics requires high-resolution methods for which access to a clean room is advisable, but that is needed only once. After fabrication of the master mold is complete, fabrication of the PDMS replica is very straightforward, albeit labor-intensive: it entails mixing two liquid reagents and pouring the mixture (the PDMS liquid precursor) over the mold. After waiting for the precursor to cure for a few hours, the researcher manually peels the flexible PDMS replica from the master (see figure 1.17 in chapter 1). The master is then ready for another replica.

Because it is so compatible with live cells and can be micromolded with ease, PDMS attracted the attention of many early microfluidic pioneers. PDMS is gas-permeable, so engineers use it to equilibrate the adequate oxygen and CO_2 concentrations that are necessary for cell survival through the walls of PDMS chambers (figure 5.1). The introduction of PDMS signified the end of clean room dependency (researchers pay steep fees to use clean rooms) and gradually moved microfluidic fabrication to individual labs. In our political parlance, we could say that PDMS catalyzed the end of elitist clean room facilities and the blossoming of a new era characterized by the democratization of microfluidics among engineers.

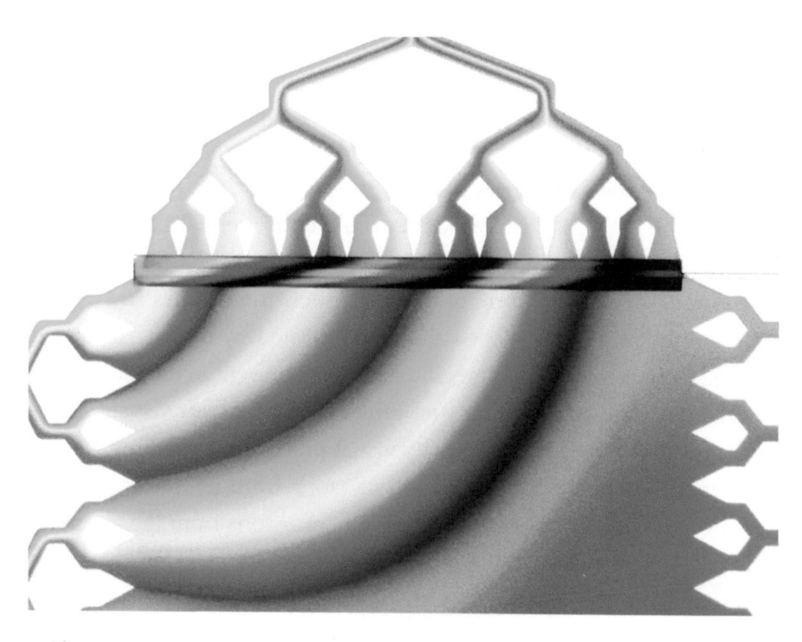

Figure 5.1
PDMS is an incredibly powerful material that facilitates the rapid, facile prototyping of cell-compatible microfluidic systems featuring spatiotemporal gradients of biochemicals, forces, and energy fields. However, PDMS also has disadvantages. *Image courtesy of Greg Cooksey and Albert Folch, University of Washington.*

CHAPTER 5

THE STRIPE ASSAY

As humbling as it may be to many of us engineers, one of the first and perhaps one of the most successful designs in the history of microfluidics was conceived not by engineers but by a group of biologists working at the Max Planck Institute for Developmental Biology in Tübingen in the 1980s.

In the early 1980s, Friedrich Bonhoeffer (box 5.1) became interested in studying how neurons are able to grow axons through long distances and find their target cells during development. Together with his technician Julita Huf, he used retina explants of chick embryos and placed them on top of a cell culture in 1982. They offered the axons two choices of growth with a glass coverslip placed at an angle. The axons could grow either on the plane of the microscope or on the coverslip, out of the plane, so Bonhoeffer and Huf could quantify which substrate was more permissive.[1] In 1987—even before Manz had made glass microfluidic channels—Bonhoeffer and his student Jochen Walter built a microfluidic device that offered axons the two choices of axon-attractive and axon-repellent paths on the same flat glass surface, so they could entice retinal axons to grow along those paths and observe them on the microscope's field of view.[2] The device consisted of a set of microchannels molded in PDMS from a photolithographically etched master fabricated by technician Bernd Stolze. They tested whether specific molecules present in the membrane of cells from the optic tectum—an important part of our visual system in the brain involved in preliminary visual processing and directing eye movement—could guide the growth of axons from the retina—which connects to the tectum—during development. Axons from the temporal side (closer to the ears) of the retina showed a preference for growth on membranes of the anterior tectum (their natural target area) over the posterior tectum. In contrast, axons from the nasal side did not show a preference. Bonhoeffer's micropatterning technique was later dubbed the *stripe assay*. It was the first PDMS microfluidic device in history.

The device was inventive even by today's standards. Atop the PDMS microchannels, Bonhoeffer placed a porous filter with a pore size of 0.1

BOX 5.1: BIOGRAPHY
Friedrich Bonhoeffer

Friedrich Bonhoeffer's life and past have revolved around the University of Tübingen, established in 1477. Tübingen is a small, old university town in southwest Germany with a reputation for having both the youngest average population and the highest quality of life of all cities in Germany. Notable Tübingen residents and scholars include the astronomer Johannes Kepler, the poet Friedrich Hölderlin, the philosopher Georg Wilhelm Friedrich Hegel, and the novelist Hermann Hesse—so excellence is mundane here.

Friedrich Bonhoeffer is an ingenious physicist turned developmental biologist who moved to Tübingen in 1960 after a postdoc at UC Berkeley. Bonhoeffer, 87 years old when I spoke to him in 2020, opened a window of his computer to his old library with the help of his son Tobias—same eyes and smile as his father—and nonchalantly explained to me that his lineage "can be traced back to the 1500s." His grandfather Karl Bonhoeffer studied at the University of Tübingen and became a notable psychiatrist and neurologist in Berlin; he had eight children, all educated at home. Two of Karl's sons, Dietrich and Klaus, became resistance fighters against the Nazi regime. Accused of plotting a coup to assassinate Hitler, they were executed in the last days of World War II. Friedrich's father, Karl-Friedrich, was a renowned chemist who had studied with Nobel laureate Walther Nernst and became the director of the Max Planck Institute for Physical Chemistry in Göttingen, also known as the Karl-Friedrich Bonhoeffer Institute. Friedrich's son Tobias studied physics at the University of Tübingen and is now the director of the Max Planck Institute for Neurobiology in Munich. The Bonhoeffers have a lot to be proud of.

microns. "We had been working with these filters for quite a long time in the DNA replication field. We used them to select mutants that are unable to replicate at a higher temperature," recalls Bonhoeffer. When they placed a solution of the cell membrane debris on top of the device and applied suction through the microchannels, the resulting flow immobilized the debris particles atop the surface of the filter, clogging it. Clogging the filter resulted in the formation of membrane fragment stripes having the same shape as the microchannels underneath the porous filter. Next, they moved the filter on top of another matrix with homogeneous porosity and added another cell extract on top of the filter. When they restarted suction, the flow immobilized the membrane fragments from this new cell extract in the filter areas between the old stripes, forming stripes of two different cell extracts. This strategy for trapping particles or cells in solution by passing them through a sieving obstacle is now termed hydrodynamic trapping, and the filters are used in numerous organ-on-a-chip designs (chapter 6).

Neurobiologists used the stripe assay for many years for the very specific use of enticing neurons to grow their axons in a straight line.[3] These scientists did not seem to realize or care that PDMS microfluidics had an immense potential for many other applications such as diagnostics, regenerative medicine, and materials synthesis. Conversely, if we are to trust the citation records, microfluidic engineers—starting with myself—were blindly unaware for more than a decade that a whole community of axon guidance researchers had been using a PDMS microfluidic device.

THE ADVENT OF SOFT LITHOGRAPHY

George Whitesides—the most cited chemist alive—became interested in PDMS from a different angle. His lab at Harvard University had been working for more than ten years on a class of organic molecules that, upon binding on one end on the surface of certain materials, spontaneously stand up like the bristles of a brush to form densely packed layers that are one-molecule thick. Scientists call these layers *self-assembled monolayers*, colloquially referred to as *SAMs*. One of the properties that tickled the fancy of the Whitesides group was that the composition of the SAMs drastically affected their wettability, as seen by the sizes and shapes of droplets forming on the surface.

At the beginning of 1988, Manoj Chaudhury, a visiting scientist from Dow Corning, the manufacturer of PDMS, "brought an understanding to the group of the properties and chemistry of siloxanes that he was using to study fundamental questions in adhesiveness," says Hans Biebuyck, who at the time was in the second year of his PhD in the Whitesides group. Siloxane is the name of the chemical group Si-O-Si, which forms the backbone of all silicones in a repeating unit. The variant with two methyl groups, poly(*dimethyl*siloxane), or PDMS, is one of the simplest of all silicones.

After obtaining his PhD, Manoj Chaudhury went to work for Dow Corning to work on adhesives, coatings, and polymers. "This is when I learned silicone chemistry and worked on the joint roles of the mechanics and chemistry of PDMS elastomers," says Chaudhury. As he was developing methods to fabricate PDMS elastomers of different shapes for his studies on contact mechanics and investigating plasma-assisted bonding of PDMS, he received an internal Dow Chemical award that allowed him to spend one year at a laboratory of his choice. He chose the laboratory of George Whitesides. "Later, when I joined the faculty of Lehigh [University] in 1993, in their infinite generosity, Dow Corning formed a chaired professorship for me and supported three students and me for about seven years," remarks Chaudhury of the Dow Corning munificence that has had such an impact in his life. One of the projects in the Whitesides lab, published in *Langmuir* in 1991, consisted of measuring the adhesion between PDMS microlenses and a PDMS surface, which were modified by oxygen plasma as well as SAMs exhibiting different functionalities.[4] Between 1991 and 1993, Chaudhury published five papers with Whitesides[5]—including three *Science* papers in ten months; four of those five papers were studies of the mechanics and the interfacial properties of PDMS.

At the end of 1991, just after Chaudhury's departure, Amit Kumar, a new postdoc (box 5.2), arrived from Caltech and started applying MEMS technology to PDMS. At the time, other members of the large Whitesides group—which comprised fifty-two members, twenty-five of whom were postdocs—had started looking into patterning alkanethiols on metals to make arrays of droplets. In 1992, postdoc Nick Abbott created a pattern of unusual square drops by micromachining a hydrophilic SAM-covered surface and depositing a hydrophobic SAM on the machined lines (figure 5.2).[6]

BOX 5.2: BIOGRAPHY
Amit Kumar

When Amit Kumar left India, he was a little boy unaware of PDMS and SAMs. He was born in a highly educated family in Patna, the capital and largest city of Bihar in northeastern India. His father had a PhD in materials science and his mother was an obstetrician-gynecologist. The family moved to the US with three children when Amit, the oldest, was seven. They then lived in various places until they settled in Southern California. Kumar was accepted into a dual-degree program between Caltech and Occidental College—a private college in Los Angeles with renowned alumni such as President Barack Obama—where he studied mathematics and chemistry. In part, because he had not enjoyed his Caltech classes, Kumar decided to go to Stanford for grad school. Ironically, his PhD advisor Nathan Lewis ended up moving to Caltech during his PhD. For his PhD, Kumar worked on photovoltaic solar cells and did a lot of electrochemistry and semiconductor lithography. He finished his PhD at the end of 1991 and, intending to exploit SAMs for nanoscale patterning, started a postdoc at the Whitesides group, the world leader in SAMs at the time.

In January of that year, Kumar started scratching SAM surfaces with pens and scalpels with the hopes of being able to deposit tiny patterns of conducting polymers that might have novel properties. Instead, the project turned into a SAM-masked, gold-etching project. Kumar, working with Nick Abbott and grad student Hans Biebuyck, realized that, once subtractively patterned with various techniques, the alkanethiols protected the metals from a cyanide etch.[7] Then Kumar decided to start experimenting with printing as a way to additively and inexpensively pattern the SAMs. First, he tried with elastomeric stamps like the types used for conveying ink to paper, and several other materials like polyurethanes, which proved useful but not ideal for high-resolution patterning. Then the eureka moment came to him.

In March 1992, as Kumar went into the lab's stockroom, he saw the box containing the two bottles for making PDMS, probably left by Chaudhury: the bottle of polymer precursor next to the bottle of catalyst. Kumar was familiar with this flexible, transparent polymer. Although he had never met Chaudhury, he had read the *Langmuir* paper in which Chaudhury had

Figure 5.2

These 4-millimeter-wide blue and green drops have been deposited on a hydrophilic self-assembled monolayer (SAM). The drops are separated by 1-micrometer-wide lines that have been mechanically machined into the underlying gold surface; the lines are covered with a hydrophobic SAM so as to repel the drops. This image, designed by Felice Frankel, was used for the cover of the September 4, 1992, issue of *Science* magazine.

Copyright: Felice Frankel, MIT.

used PDMS contact lenses. Kumar knew that upon mixing the prepolymer with the catalyst at 10:1 ratio and mild heating, the concoction slowly solidifies into crystal-clear, rubbery solid. Kumar's ink stamps were elastomers and had not worked quite right, but maybe—he surmised—the PDMS elastomer might work better. He still had a photolithographically fabricated wafer and decided he did not risk much by pouring a bit of PDMS over it and letting it cure overnight.

When Kumar came back the next day, he carefully peeled the PDMS off. "It was incredibly colorful because the [PDMS replica] pattern was diffracting light," Kumar recalls with excitement. "I immediately put it under the microscope, and the features had been reproduced so tightly, and so sharply—it was amazing!" As soon as Kumar had the PDMS replica in his hand, he realized he had created something unique. Unlike all the diffraction patterns he had seen until then, this one was transparent and flexible—an idea he kept in mind for later.

He next used the PDMS pattern to stamp an alkanethiol SAM. When he etched the stamped SAM with cyanide—a ten-minute experiment—he saw that the raised stamp features faithfully reproduced on the gold surface while the rest—unprotected by the SAM—had been etched by the cyanide. Kumar initially called the technique rubber stamping, but Whitesides decided the name microcontact printing was more appropriate. After some optimization, they published this simple technique in 1993 in *Applied Physics Letters*,[8] a paper that has been cited more than 2,400 times as of 2021.

Kumar's technique formally introduced PDMS to the microfabrication community and ignited George Whitesides's imagination. Whitesides was aware of MEMS technology in the early 1990s, and Kumar gives credit to Whitesides for grasping the big picture even better than himself. Whitesides clearly heralded the simplicity of PDMS molding (dubbed soft lithography) and had high hopes for replacing photolithography as the standard research tool for micropatterning materials:[9] scientists manually mixed and poured the PDMS precursors into a mold to form 2D-layer replicas, and aligned and bonded the mold replicas by hand to form the final, optically clear device.

The last month before Kumar left for a job in industry in 1993, Whitesides asked him one last, rather unusual request: to please write all the ideas that he had thought of during his time developing PDMS rubber stamping, so they would not get lost in research nirvana. Kumar agreed and wrote a several-page list of ideas that included using the flexibility of the stamps for patterning curved surfaces (developed by Jackman et al.

into two *Science* papers in 1995 and 1998[10]) or for deforming diffracting elements to make adaptive elastomeric optics.[11] As a reward, Whitesides later included Kumar in several relevant patents.

A PERSONAL VIEW

I witnessed the baby steps of soft lithography and PDMS microfluidics from a nearby post because I had started a postdoc at the MIT Department of Chemistry with Mark Wrighton in 1994 to pursue a project that entailed fabricating microcantilevers.[12] I did not know any microfabrication at the onset, but fortunately my co-advisor was Marty Schmidt, the MEMS expert in MIT's Electrical Engineering and Computer Science Department. I say "fortunately" also at a personal level—Marty is a wonderful, down-to-earth fellow. "I'm a gadget guy," Marty told me once with his Santa Claus smile to summarize his contagiously playful approach to research. At the time, Whitesides was publishing his first PDMS papers 1 mile down the road from Massachusetts Avenue. In 1995, Enoch Kim, another Whitesides student, demonstrated the use of PDMS devices for micromolding polymer precursors in capillaries (MIMIC), essentially a microfluidic patterning technique; the technique was awarded a cover in the journal *Nature*.[13] At MIT, my building—the Microsystems Technology Laboratory, Building 39—was plastered with posters of silicon and glass devices, so Kumar's rubber stamps and Kim's fluid patterning technique seemed, at least intuitively, very transformative: they could add new materials to the old MEMS toolset.

Mark Wrighton was good at giving concise but useful advice, so I asked him in 1995 if he knew something about the new stamping technique. "George is a good friend. Go see him," Wrighton answered in his usual curt style. My visit to the Whitesides lab was very brief. There was a bit of confusion at first because Whitesides asked me where my biosketch was; he thought I was applying for a position. When I clarified that I worked with Mark Wrighton and Marty Schmidt, he said with a witty smile, "Why is Marty interested in a technology that will put him out of business?" I had to laugh because it was just me, not Marty, who was interested. In our many encounters later, Whitesides would show that he had a cunning sense of humor, so I have always suspected he did not mean that seriously—but I was not in a position to challenge his point of view anyway. After my second clarification, Whitesides quickly proposed going to the lab to introduce me to one of his students so I could see for myself. PhD student Xiao-Mei Zhao taught me how to mix PDMS and microstamp alkanethiols in about ten minutes. For a physicist-turned-engineer, the simplicity of it was both disappointing and refreshing. I was immediately persuaded that this was much easier than gowning up in the MIT clean room.

Marty Schmidt, true to his love for gadgetry, encouraged my first experiments with PDMS.[14] I loved watching how PDMS, due to its surface free energy, establishes a conformal seal with smooth, dry surfaces, and how fluids penetrate the channels with polymer precursors under the microscope. I also loved working for Marty, an unpretentious man who has always reminded me of good bread—a warm heart surrounded by a crisp mind. But my fellowship was coming dangerously close to an end. Toward the end of 1996, Mehmet Toner was pursuing a project to build a bioartificial liver when he decided he needed a postdoctoral MEMS engineer for the project and posted an ad. Marty, who knew Mehmet Toner, alerted me about it. That might have been one of the luckiest days of my life. I remember being impressed by the name in the ad: Center for Engineering in Medicine. I didn't even know what side of the body the liver was on back then, but fortunately I was not quizzed on that during my interview, and I was hired to start on January 1, 1997.

The bioartificial liver project envisioned the scale-up of a liver cell culture using a set of stacked cell culture substrates in microfluidic chambers. Conceptually, it was the inverse of the problem solved by Andreas Manz, who had miniaturized chemical reactors into a microTAS (chapter 3). It was clear from the start that the scale-up would be fraught with challenges. A significant problem was the cells themselves: liver cells are not happy when dissociated into single cells and placed on a plastic surface outside of the soft, blood-bathed environment of a real liver. Although the group had done great work to improve the cells' viability and function, we did not know how the cells would perform under flow or in a microchamber.

Somehow it must have been obvious that I was more interested in tinkering with PDMS on the side. "A good scientist doesn't try to prove his hypothesis right but his hypothesis wrong," Toner had lectured me, enigmatically, one day. At first, Toner respected my passion for PDMS at a distance, but after a few months, he told me to stop working on the bioartificial liver and joined my PDMS games like my best friend. Microfluidics can be infectious.

THE MICROFLUIDIC IMMUNOASSAY

The microfluidic immunoassay was born out of research on monolayers at IBM Zurich in the 1990s. IBM Zurich had received two consecutive Nobel Prizes (one for scanning tunneling microscopy [STM] in 1986 and one for high-temperature superconductivity in ceramic materials in 1987), so it had become the mecca of many young academics itching to explore new materials frontiers.

A French student at Toulouse named Emmanuel Delamarche was one of them. While Delamarche was doing his masters on SAMs on gold surfaces and studying them using STM under Christian Joachim, he learned through Joachim that IBM Zurich was looking for a PhD student. Delamarche applied and got the position. The weekend after he had finished his masters, Delamarche packed things in the car and drove to Zurich. "Half of my belongings were papers from Whitesides and literature on SAMs," says Delamarche. It was the summer of 1992.

Delamarche joined the team of biochemist Bruno Michel at IBM Zurich. Michel was involved a few years earlier in Rohrer's pioneering experiments with STM on SAMs[15] and was now working with a cantilevered STM tip to measure the forces applied to SAMs during STM imaging.[16] In 1994, Hans Biebuyck joined the team. Biebuyck had accumulated a wealth of knowledge on surface science during his eight-year-long PhD in chemistry at Harvard working on SAMs, but—as time would prove—had not given the best of his creativity yet. Delamarche describes Biebuyck as a "living Wikipedia before the internet era." Importantly, Michel was quite open to trying new things. "At a time when we were very biased by lithography, Hans introduced us to microcontact printing," remembers Delamarche. Michel welcomed Biebuyck's idea of focusing on microcontact printing and a project—internally dubbed microcontact processing—was officially started around 1996 to develop micropatterning techniques for microelectronics and protein biosensors.

The group started molding PDMS with what Delamarche recalls being "boring patterns." A smart engineer on the team, Heinz Schmid, molded more creative features like an eye, a bee, and the M of the IBM logo. They had fun filling the molded eye, bee, and M under the microscope with different solutions of ethanol containing various dyes. The critical moment was in September of 1996 when Biebuyck entered the lab, looked at the experiment, and—obviously unimpressed—told Delamarche: "hmm . . . maybe it could be interesting to fill the structures with solutions of proteins," and left the room. That remark prompted Delamarche right away to mold some channels linked to reservoirs, make the PDMS hydrophilic using an air ionization treatment, and draw solutions of fluorescently labeled antibodies through the channels by capillarity. "The antibodies formed patterns on the surfaces with surprising ease and resolution. We immediately thought of miniaturizing immunoassays," Delamarche remembers.

The considerable cost of antibodies (box 5.3) ignited Delamarche and Biebuyck's idea of miniaturizing immunoassays: if they could perform an immunoassay with one hundredth less total volume of reagent—at the same usual concentration—then the cost of the assay would be reduced by the same factor. They submitted the paper to *Science* in December 1996, and it was accepted in mid-1997.[17] I vividly remember the day when my postdoc advisor Mehmet Toner came to my office to show me this paper

BOX 5.3
Immunoassays

Immunoassays are measurements that use antibodies, the molecules produced by cells of the immune system to neutralize pathogens. Scientists have learned to purify antibodies because they can be handy tools for identifying biomolecules in biotechnology and biomedical research. The first such measurement was devised in 1959 by Rosalyn Yalow and Solomon Berson;[18] they purified antibodies against insulin to measure the concentration of insulin in human blood. Any substance that induces the immune system to produce antibodies against it is called an antigen. Yalow and Berson could apply the assay to the measurement of virtually any biomolecule (acting as the antigen) that they could inject into a patient or an animal to produce antibodies against it. In 1975, Georges Köhler and César Milstein were able to produce antibodies from continuous "hybridoma" cultures of fused cells, bypassing the need for animals and producing "monoclonal" antibodies that bind with high specificity to their antigen.[19] The importance of these techniques in biology and medicine cannot be overemphasized. Yalow (after Berson's death) received a Nobel Prize in 1977, and so did Köhler and Milstein in 1984. During the next decade, pharmaceutical companies raced to produce antibodies for every imaginable biomolecule—since then, the production methods have not changed significantly, and antibodies remain an expensive tool (on the order of hundreds of dollars for a few micrograms).

and how excited we got reading it—and I'm sure many others around the world were sharing our excitement.

THE PDMS EXPLOSION: CELLS, 3D, AND AUTOMATION

Biebuyck's *Science* paper illuminated the way for many of us who wanted to use PDMS for biomedical applications. In 1997, just four months after Widmer died, Carlo Effenhauser—who, like Manz, had worked for Widmer and stayed at Ciba-Geigy (now Novartis)—used a PDMS device to separate and detect DNA molecules by capillary electrophoresis with single-molecule resolution.[20] In 1998, Toner and I demonstrated PDMS microfluidic devices for cell and protein micropatterning.[21]

However, the toolset was still very rudimentary. At a group meeting, Mehmet Toner expressed his frustration about these "Mickey Mouse gadgets" and called for more user-friendly devices. A particularly aggravating problem was that the PDMS channels were not strongly bonded to the surface, which meant that they would detach as you injected fluids with too much pressure into the channel. You did not know at what pressure that would happen.

George Whitesides left a copy of the Effenhauser paper on the desk of his postdoc David Duffy and asked if they could make a capillary electrophoresis device. That got Duffy talking to Olivier Schueller, another postdoc, about microfluidics. "We tried the Effenhauser way, but untreated PDMS didn't cut it in terms of bonding (it leaked) or charge properties (it didn't have a charge) to allow electrophoresis," remembers Duffy. Those experiments led Olivier to investigate the oxidation of PDMS surfaces with an oxygen plasma. A plasma is a gas subjected to high voltages to generate highly reactive ions, in this case oxygen ions. "Since we were in a surface chemistry group, we were always trying to find ways to modify surface properties," recalls Schueller. Chaudhury's experience with PDMS at Dow Corning was also helpful. They tried different things, different pressures, and durations. "Durations turned out to be a big deal. Too long in the plasma chamber, and it did not work. The duration was always key. There was a 'window' for bonding!" remembers Schueller of the critical moment. They used either glass slides as substrates or silicon wafers and, at some point, a piece of PDMS adhered to the glass. "The ability to bind PDMS to glass or silicon was awesome. Then we also tried PDMS to PDMS. And we kept on trying things," adds Schueller. Because the bonding had introduced surface charges, "it also helped support capillary electrophoresis. From that, it all fell into place," summarizes Duffy. Duffy, Schueller, and grad student Cooper McDonald coauthored a paper in 1998 that introduced the ability to seal channels irreversibly using an oxygen plasma.[22] The paper, titled "Rapid Prototyping of Microfluidic Systems in Poly(dimethylsiloxane)," is by far the most cited paper in the history of microfluidics at almost 6,400 citations by 2021. It has been useful to many people.

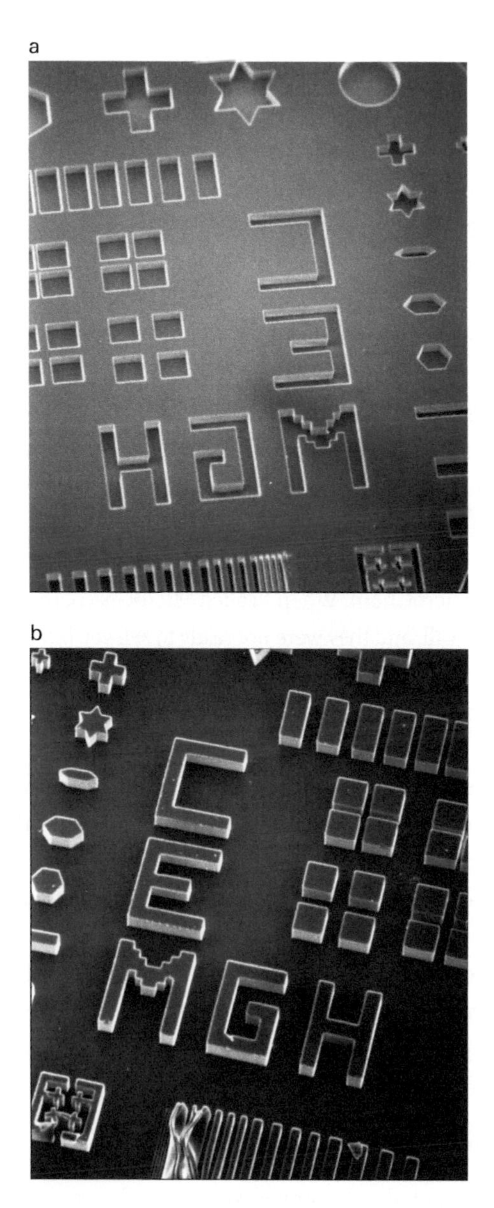

Figure 5.3a and Figure 5.3b

(a) SU-8 photoresist micropattern and (b) its PDMS replica. The depth/height of the pattern is 53 microns. The straight sidewalls that can be achieved by SU-8 have allowed for producing rectangular-profile microfluidic channels in PDMS. *Images by Albert Folch, then at Center of Engineering in Medicine, Harvard Medical School.*

The Duffy et al. paper also described the first use in microfluidics of the negative photoresist SU-8, a photoresist that allows for creating vertical sidewalls and tall features (figure 5.3). SU-8 had been reported by IBM in 1995[23] but did not become commercially available until two or three years later. MicroChem, a small company in Massachusetts, commercialized it. "We were extremely interested in finding ways to make thicker structures than what was achievable with Shipley positive resists, which was our go-to until then," recalls Schueller. "We also visited MicroChem once, as they were local. Nice folks there. Friendly. Willing to find a way to expand the business somehow." I remember this point in time vividly as well. I had stumbled upon the IBM paper in 1997 before Duffy et al.'s paper came out and immediately called the IBM group from my desk at the Toner lab. They referred me to MicroChem. When I called MicroChem, they said I was the first customer to call, and they were not ready to sell yet, but they were happy to send me a bottle for free. Deal! We found that SU-8 greatly facilitated the process of mold fabrication and PDMS de-molding.[24] Just as important, SU-8 provided a tool for making rectangular-profile channels where one could model the flow resistance of every segment with a simple formula. We could now calculate the flow rates in a complex network with Excel.

The year 2000 was important both for the field and for me personally. I started at the University of Washington in June 2000, after marrying the beautiful biologist that I had followed all the way to Boston. Shuichi Takayama, who goes by Shu and was then a postdoc in the Whitesides lab, had admirably demonstrated a few months before how he could use two flows to squeeze a central flow and pattern the fluidic environment of cells dynamically using PDMS microchannels.[25] By varying the pressures on each inlet, it is possible to steer the pattern of the fluids over the cells (figure 5.4), as if pulling on a bridle to steer a horse. Paul Kenis et al. (Whitesides) had also illustrated vividly on the cover of *Science* how they could harness multiphase laminar flows for patterning surfaces.[26] Stephan Dertinger in the Whitesides group designed a gradient generator that looked like a Christmas tree[27]—the branches served to split and recombine the streams (figure 5.5)—and, with Daniel Chiu and Noo Li

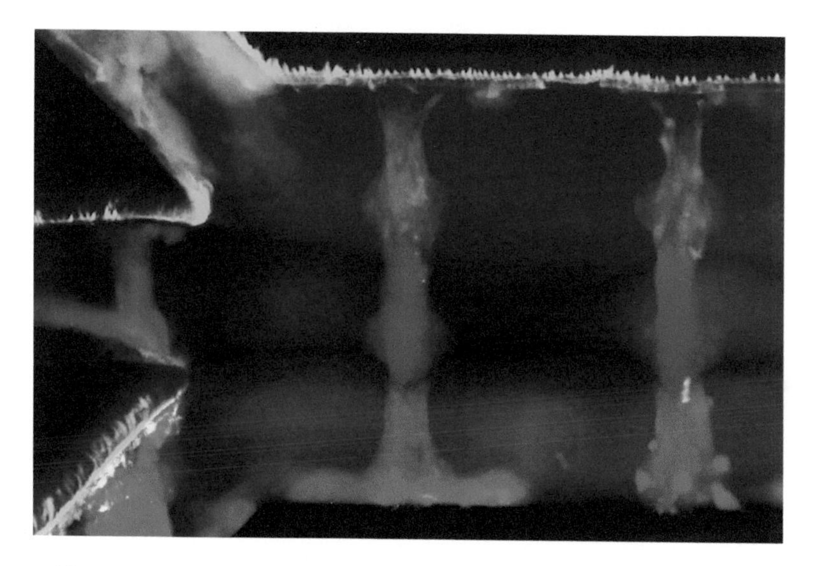

Figure 5.4
Live, multinucleated muscle cells have been grown along adhesive patterns that span the width of a microfluidic channel at regular intervals. The three inlets of the device are used to expose different parts of the cells to a laminar flow of different dyes (green, blue, and red). Flow is from left to right.
Image courtesy of Anna Tourovskaia and Albert Folch, University of Washington.

Jeon, illustrated how the device could guide the growth of neurons[28] and the migration of neutrophils.[29] This work was followed by many others and unleashed a frenzy to make better gradient generators,[30] for example featuring arrays of gradients, with minimal flow, and with pipette access (no roof, see figure 5.6). Laminar flow has existed forever, but PDMS microfluidics gave us the first LEGO-like tool to manipulate it. If PDMS provided the raw canvas for microfluidics, laminar flow provided the palette to generate multicolor paintings.

Until this point, PDMS microfluidic devices were two-dimensional, which limited the complexity of the fluidic patterns that researchers could achieve. In 2000, both Janelle Anderson et al. (Whitesides)[31] and Beebe's group[32] demonstrated the multilayer fabrication and assembly of 3D

Figure 5.5
Gradient generator. Flow is from top to bottom. The blue dye mixes with the yellow dye via the principle of splitting and recombining. Toward the middle of the device, all the streams are joined into the observation chamber. A third inlet is used at that point to introduce cells into the gradient.
Image courtesy of Arul Jayaraman, Texas A&M University.

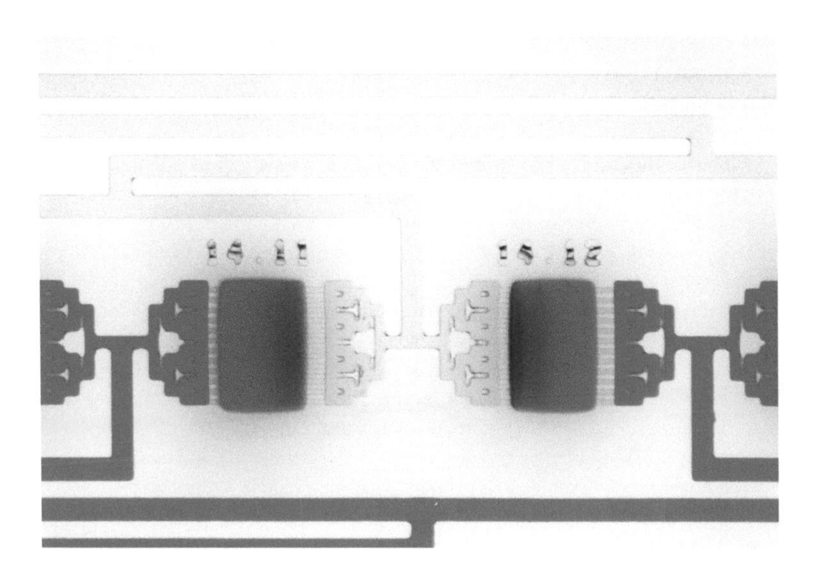

Figure 5.6
Gradient generator based on injecting minimal amounts of flow (shown in red and blue) into an open-air chamber. The device consists of 1,024 of these chambers operating in parallel; this image is a close-up, showing only two of the chambers.
Image courtesy of Nirveek Bhattacharjee and Albert Folch, University of Washington.

networks of PDMS; Daniel Chiu et al. (Whitesides) used these 3D PDMS networks for the microfluidic micropatterning of cells and proteins.[33] Also in 2000, Steve Quake's group reported multilayer techniques for building PDMS valves (in essence, a 3D structure). This contribution made the field of microfluidics explode with automation and caused it to grow exponentially (see "The PDMS Microvalve" section).

It was an exhilarating moment because you could tell that the field was coming to a boil.

* * *

I must emphasize here the monumental size of Whitesides's legacy. He fathered a prodigious generation of microfluidic engineers—Shu

BOX 5.4
An Image Is Worth a Thousand Words

Whitesides is also well known among many laypeople because he is a mesmerizing communicator. One of the more impactful and lasting of Whitesides's contributions might be his collaboration with scientist and photographer Felice Frankel. Frankel is from Brooklyn, New York, and went to Brooklyn College, where she got her science education. "I don't call myself an artist. My purpose is to communicate science, not to communicate my feelings," she says. After college, with two kids, she volunteered as a photographer at a public TV station in Springfield, Massachusetts, and published her first book, *Modern Landscape Architecture*. In 1991, she applied for a midcareer fellowship at the Harvard School of Design and got it. But she was missing science. She started taking science courses. Someone recommended to her that she take a "very visual" course in chemistry. That was George Whitesides's class. At the end of one of the lectures, she approached him and asked if she could perhaps stop by the lab and take a look at his students' work, to which he agreed. At the time, Nick Abbott's paper had just been accepted in *Science*. "His pictures could have been better," says Frankel now, but back then, she proposed to Nick: "Can we create a sample that will really communicate what is going on?" Nick produced a grid for her to play with drops. Frankel took a mesmerizing picture of a square mosaic of blue and green drops, and *Science* accepted it for the cover (figure 5.2).

Impressed, Whitesides said, "You are on to something."

In 1994, Frankel got a position at the MIT Edgerton Center, which led to a position in the School of Science. She gradually developed a friendship with Whitesides, and she proposed to him to put together a book with her pictures and his words. In 1997 they published *On the Surface of Things*, a gorgeous dance of words and microscale-phenomena photographs. Whitesides and Frankel seemed to amplify each other's talents for the enjoyment of the community. In 1999, Paul Kenis had demonstrated the use of the laminar flow technique, a manuscript that had been accepted to *Science*.[34] However, the figures only displayed two flows in an unappealing zig-zag pattern. Felice intervened again: "We can do better. Can you make a PDMS channel with this design"—here she drew seven channels—"and fill it with different colors?" She asked for seven channels because the *Science* logo has seven letters. Kenis made the pattern of seven channels, and Frankel photographed it under the microscope (figure 5.7). The art director of *Science* picked up on the idea and created a logo with seven matching colors. The resulting cover of *Science* is now iconic.

Figure 5.7
Seven different color dyes are introduced into a microfluidic channel through different inlets, resulting in a parallel laminar flow of seven streams. Flow is from top to bottom. This image was designed by Felice Frankel and used for the cover of *Science* magazine in the July 2, 1999, issue (box 5.4).
Copyright: Felice Frankel, MIT.

Takayama, Daniel Chiu, John Rogers, Rustem Ismagilov, Noo Li Jeon, Sam Sia, Abe Stroock, and many others—that sprung out of his lab in a few years and continue to spread the microfluidic gospel in academia to this day. Had it not been for Whitesides, microfluidics would not be half the field it is today. And if you have ever sat down with him at lunch and witnessed his witty dialectic abilities—once in Heidelberg, I saw him persuade a bunch of sophisticated Europeans (who ignored he was born in Louisville, Kentucky) that the Kentucky Fried Chicken founder was a culinary genius—you'd agree.

Yet after PDMS brought about the new democratic access to microfluidics for researchers, there was still another—previously unforeseen—power subjugating them. As devices became more complex, flow control became increasingly limiting. Biologists and chemists progressively relied on gathering more and more data points but were subjected to "the tyranny of pipetting"[35]—and many still are (figure 5.8). Pipettes are widely used tools in chemistry and biology laboratories to measure and dispense volumes of liquids, but pipetting for hours can become tedious and stressful. Perhaps just as important, the most precise pipettes cannot dispense cellular-scale fluid volumes so the reactions are inevitably scaled up, which consumes precious reagents. In the late 1990s, you might have also turned your attention to the microfluidics field for a solution to this problem, hoping to miniaturize and automate fluid handling to reduce both the amounts of human labor and supplies. But until the year 2000, the equivalent of a mechanical valve or a rotary pump was missing, so microfluidic engineers lacked a practical tool to switch flows on and off.

Until Marc Unger, Stephen Quake, and coworkers, then at Caltech, published in the April 2000 issue of *Science* the PDMS microvalve.[36]

THE PDMS MICROVALVE

Others had previously directed flow with voltages, or made valves using a PDMS membrane and a rigid seat, but this valve was entirely made of PDMS, and could be fully integrated with the rest of the device. Marc

Figure 5.8
Biologists and chemists use an instrument called a pipette to transfer fluids from one container to another. Though simple, this repetitive procedure can become tedious and lead to stress and fatigue. Even the most accurate pipettes dispense amounts of fluids that are orders of magnitude larger than the volume of a single cell. The PDMS microfluidic valve sought to drastically reduce both the labor and the amount of fluids involved in fluid manipulations that are central to most chemical and biochemical reactions.
Image courtesy of the University of Washington.

Unger, originally from New York State, was attracted to Caltech because "it had the highest 'genius density' of all the schools" he visited. For his PhD in physical chemistry in the lab of John Baldeschwieler, Unger put a single molecule on the end of an atomic force microscope tip. Near the end of 1997, Steve Quake came to visit the Baldeschwieler lab looking for a postdoc for his human genome project. The idea was to sequence DNA by tethering it in a microchannel, giving it a primer and a polymerase, and feeding in a single nucleotide base at a time. They could fluorescently label a percentage of the bases, and then they could shut off the fluorescence by intense illumination (a technique called photobleaching). "It sounded like a cool project, and Steve was clearly a very smart guy," recalls Unger.

There was only one problem, according to Unger: he was not done with his PhD, and Quake's grant had to get moving, or it would be withdrawn. Unger then decided to take a big risk. He took a leave of absence from his PhD to pursue Quake's project.

Marc Unger started working on the DNA sequencing project in the Quake lab on January 1, 1998. (He later would take a leave of absence from the Quake lab to finish his PhD in June 1999.) The lab was already using PDMS molding, and Unger planned to apply PDMS microchannels to the challenge. As a chemist, he became interested in using the chemistry of PDMS to anchor the DNA that they wanted to sequence. Unger noticed that PDMS was a two-component mixture: the A component had vinyl groups ($-CH=CH_2$); the B component had silicon hydride groups (-Si-H). These two groups crosslink to form an elastomer. Usually, one mixes the two components at a 10:1 ratio. Unger assumed that mixing the components off ratio would yield excess vinyl if he made the mixture A-rich, and excess Si-H if he made it B-rich, but trying to attach DNA that way was an abysmal failure. Unger wound up building the DNA attachment chemistry on glass, but that first failure proved to be providential later.

To do sequencing by synthesis, Unger would need to switch the flow of the different nucleotide precursors A, C, G, and T in a preprogrammed order. Unger soon identified the obstacle he was facing: the smallest commercially available valve he could find had a dead volume of 16 microliters, much too large. At the flow rates he could reasonably get with the microfluidic channels they were using, the characteristic time for exchanging the fluid was on the order of half an hour—clearly far too slow to be practical. Unger needed a valve with a much smaller dead volume. There was already someone in the lab who was trying to make a PDMS valve, but "they were doing something that looked horribly complicated, using magnets—that was when the insight struck," recalls Unger. "I realized that I could use the off-ratio bonding to make a crossed-channel architecture: if I made a thin layer with a channel, and then bonded a thick layer with a channel on top of it, then applying pressure to the upper channel would close the lower channel!" It worked the first or second time. But then, for some reason, the valves stopped working.

A key element turned out to be the shape of the flow channel (the lower channel). Hou-Pu Chou, a Taiwanese PhD student in the lab who had an undergraduate degree in physics and a background in MEMS, brought the necessary insight and helped make the molds. To start, they were making the molds by KOH etching of silicon, which produces the famous trapezoidal cross section—which was so useful to Kuhn and Bassous for making the first MEMS-based inkjet printer nozzle (chapter 2)—and, as a result, did not close properly. "We were scratching our heads trying to figure out why the new valves weren't closing," says Unger. Finally, staring at the surface profilometry data, Unger realized that the initial molds—the ones that had worked—were "just not quite as nice": the corners on the trapezoid were rounded. Unger hypothesized that round would work better, which is logical given the geometry of the valve deflection. But how could rounded channels be fabricated?

Chou went into the lab and came back in less than twenty-four hours with molds made by photoresist reflow, a technique invented a decade earlier to make microlenses.[37] Chou had created photoresist lines with square profiles, heated them to ~140°C to melt the photoresist into a half-pipe shape, and cooled down the wafer to obtain rounded, solid photoresist lines. Unger remembers Chou's contribution as "amazing."

The hypothesis turned out to be correct—the valves made from those molds worked flawlessly. They created the first valve on May 14, 1999. A few days later, they figured out how to make pumps by operating three or more valves in series, alternating their activation in a 1-2-3 repetitive pattern. They submitted the paper to *Science* at the end of November, and Quake made a tongue-in-cheek bet with Unger that *Science* would reject the paper. Unger says he still has Quake's $50 bill somewhere, with the signatures of all the authors.

PDMS valves have introduced automation at nanoliter scales where the fluid properties are dramatically different from those at milliliter or larger scales. The PDMS microfluidic valves have, indeed, freed biologists from "the tyranny of pipetting"[38] by enabling the inexpensive miniaturization, automation, and high-throughput multiplexing of molecular biology protocols on the same platform.[39] Quake's group exploited this automation

BOX 5.5
PDMS Automation

PDMS valves spurred the monolithic integration of various elastomeric microactuators into microfluidic devices, such as cages to measure DNA binding,[40] reconfigurable micro-optical elements,[41] active micromixers,[42] dynamic gradient generators for single-cell analysis,[43] flow rate control elements,[44] worm traps,[45] combinatorial multiplexers,[46] biopsy assays,[47] and pneumatic microvortexers.[48] Using valves of increasing sizes, it became possible to actuate pumps with a single control line instead of three, thus saving real estate.[49]

In the Unger/Quake microvalve design, the valve is open at rest and is actuated pneumatically. Just three months after Quake submitted his manuscript to *Science*, while it was still in press, Kazuo Hosokawa and Ruytaro Maeda submitted an entirely different PDMS microvalve design, also pneumatically actuated and equally ingenious, this one closed at rest.[50] Maeda's design (figure 5.9) can be fabricated by SU-8 photolithography, so it does not require the critical step of photoresist reflow. On the other hand, the off-ratio bonding technique invented by Unger was immune to misalignment, which allowed for adding thin layer after thin layer to create complex multilayer structures with micron-scale features. With either type of valve, researchers could fashion micropumps and multiplexers (figure 5.10).

Figure 5.9
PDMS valves controlling the flow of sixteen laminar flows. Flow is from bottom to top. The top valve is open and the side valve (which diverts the flow orthogonally toward the side) is closed.
Image courtesy of Greg Cooksey and Albert Folch, University of Washington.

BOX 5.5 (continued)

Figure 5.10

A PDMS multiplexer based on pneumatic PDMS microvalves (the small buttons at the center of the image). The multiplexer allows for selecting eighty-one unique chemical combinations of sixteen inlet compounds and routes them into a chamber either as a homogenized mixture, a step gradient (as shown), or a smooth gradient.

Image courtesy of Greg Cooksey and Albert Folch, University of Washington.

Quake's group has developed devices that automate cell sorting,[51] PCR,[52] nucleic acid purification,[53] single-cell messenger RNA and complementary DNA analysis,[54] DNA sequencing by synthesis,[55] protein crystallization,[56] and long-term bacterial culture,[57] to name a few molecular and cellular biology protocols. It was not a brute-force miniaturization effort—some of the efforts went into

(continued)

BOX 5.5 (continued)

entirely new devices, such as a novel redesign of the valves for mechanically trapping DNA molecules in the midst of binding to its transcription factor partners (published in *Science* in 2007).[58] Like Whitesides, Quake has three papers among the ten most cited papers in the history of microfluidics.

capability and, in a few years, brilliantly developed the microfluidic version of every conceivable molecular biology protocol previously requiring more complex instrumentation, longer times, and more reagents than with microvalves (see box 5.5). In 2002, the Quake lab demonstrated a 1,024-valve multiplexer that was featured on the cover of *Science*. Many others have followed (figure 5.11).

The commercialization of PDMS microfluidics has not been easy, largely due to fabrication obstacles (box 5.6), but some companies have plowed through. Steve Quake and Gajus Worthington started Fluidigm in 1999 to commercialize devices with PDMS microvalve technology. In 2005, Fluidigm established a biochip manufacturing facility in Singapore. This facility produces the most complex and beautifully crafted PDMS chips in the world. Marc Unger's MSL valves (as Unger likes to call them) and his off-ratio bonding technique have enabled the production of six-layer chips. One of Fluidigm's products is the BioMark 48.48 dynamic array, a microfluidic device that performs quantitative real-time DNA amplification. Using MSL valves, the BioMark 48.48 array partitions and combines reagents and samples into an impressive 2,304 reactions. For example, each BioMark microfluidics platform can process thousands of COVID-19 PCR tests per day from saliva samples. With more than five hundred employees, Fluidigm is now arguably the largest microfluidics-based company in the world. Isoplex, a spin-off of James Heath's lab, commercializes a platform for measuring cytokines released from single cells cultured on a microfluidic array with PDMS valves.[59] Aspect Biosystems, based in Vancouver (Canada), manufactures 3D printers with

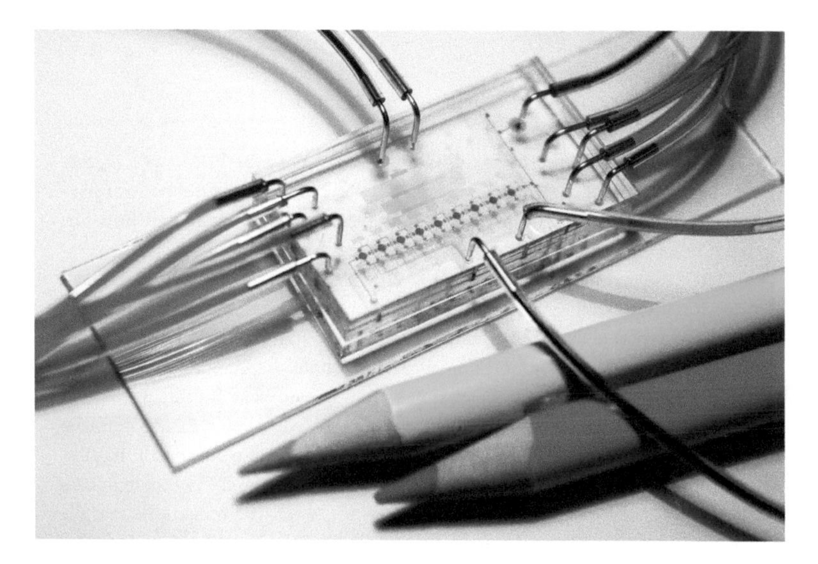

Figure 5.11
Microfluidic platform for the automated stimulation of populations of single cells. The device produces a linear gradient of nine concentrations that are delivered to an equal number of microchambers, each containing 492 microwells, where individual cells are captured. PDMS microvalves allow for isolating the microchambers and for delivering chemical waveforms of tunable frequency and amplitude.
Image courtesy of Alan González-Suárez and José Luis Garcia-Cordero, CINVESTAV-Monterrey, Mexico.

microfluidic printheads that can switch the flow of up to five inks with PDMS valves.[60]

But PDMS can also be a treacherous material (box 5.6). The fabrication process, so straightforward to implement on a small, manual scale, is very costly to automate. Perhaps even more important, proteins stick to the PDMS surface, and the PDMS walls behave like a sponge for small hydrophobic drugs—these problems can be fixed with additional coatings, but they come with additional manufacturing costs. Xona Microfluidics, a spin-off from Noo Li Jeon's lab founded in 2008, successfully sells simple (non-valve)

BOX 5.6
The Challenges of PDMS

Despite all these successes, it would be unfair to describe PDMS as a perfect material exempt of challenges. Although PDMS remains the most popular platform for microfluidics prototyping among researchers, you will rarely find it in commercially available products for several reasons.

First, the surface of PDMS is not as suitable as that of glass or thermoplastics for immobilizing biomolecules, although a fix can be arranged—for a price. As reported by Biebuyck and colleagues in 1998, the PDMS hydrophobic surface causes surface adsorption of proteins onto the channel walls, which by depletion of proteins in the channel solution can make it difficult to measure small concentrations of protein analytes.[61] Hydrophobic PDMS surfaces can be rendered hydrophilic (e.g., via an oxygen plasma treatment[62]), but the PDMS surface undergoes spontaneous hydrophobic recovery in less than one hour.[63] It is also possible to coat PDMS with a layer that renders its surface protein-repellent. However, these supplementary treatments add costs to the manufacturing of microfluidic devices and make the goal of commercializing inexpensive microfluidic devices even more challenging.

Second, PDMS appears to behave like a sponge for small hydrophobic compounds such as low-molecular-weight drugs. Again, this problem also has a fix, but at an additional cost. Chromatography studies have even used PDMS coatings to extract organic volatiles from fluids since the 1980s.[64] Beebe's group published in 2006 the first[65] of a series of landmark studies that showed that PDMS—due to its porous and hydrophobic nature—can absorb small lipophilic molecules into the bulk of PDMS microfluidic channels[66] (e.g., Nile Red, Rhodamine G, estrogen, diazepam). This PDMS porosity makes it a poor choice for building clinical devices such as drug testing and drug screening devices. Researchers have developed many coatings for PDMS to remedy, even if only partially, the problem of small-molecule absorption into PDMS, such as sol-gel[67] or silicate glass.[68] However, these modifications add substantial processing costs and time to the prototypes and modify the elasticity of PDMS. In sum, both *ab*sorption *into* PDMS and *ad*sorption *onto* PDMS can potentially alter experimental outcomes by changing the target concentrations and by partitioning molecules in undesired regions of a microfluidic device.

Third, although in the literature engineers typically describe PDMS molding as a simple procedure, it turns out to be rather labor-intensive, requiring manual steps. For more complex, multilayer devices such as organ-on-a-chip and valve devices, fabrication requires personnel with specialized training. Even though an engineer can quickly learn the technique for simple designs, more complex 3D designs necessitate fine dexterity and often result in low yields (figure 5.12). 3D printing finally helped dodge this challenge. Glick et al. elegantly bypassed alignment and simplified 3D assembly by devising a PDMS replica-molding process based on 3D-printed molds.[69]

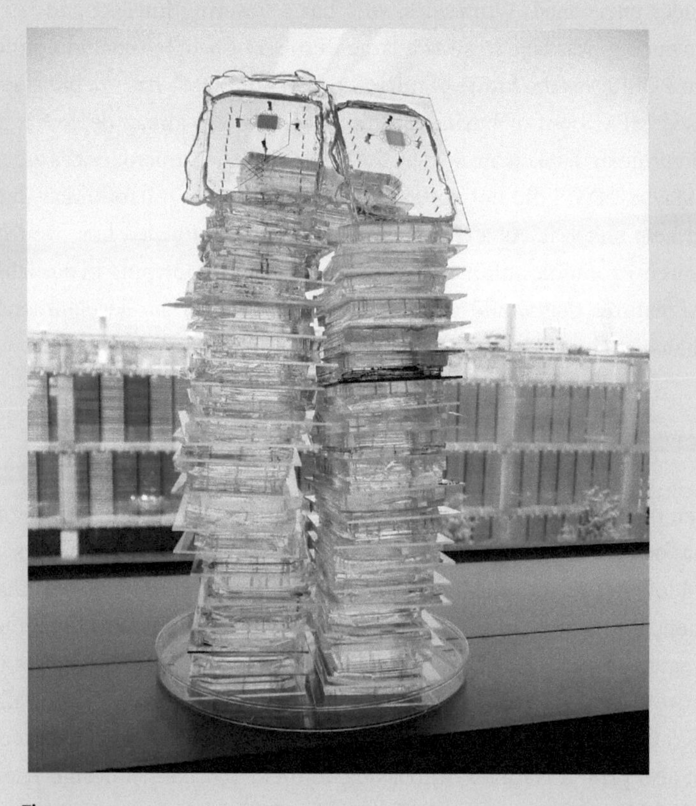

Figure 5.12
A "cemetery" of defective PDMS microfluidic devices, all created from the same molds. Each replica is composed of three layers. Low yields are not unusual in manual-based fabrication of multilayer PDMS devices.
Image courtesy of Anthony Au and Albert Folch, University of Washington.

devices for in vitro neuroscience. Initially, the company sold its devices in PDMS, but it has recently shifted to injection-molded COC devices.

For these reasons—which were not apparent in the mid-1990s—soft lithography has not been widely adopted outside research labs as Whitesides envisioned. Whitesides, who has a towering intellect and is very convincing, was able to sway a large number of people—including me—that PDMS was the future of microfluidics. But PDMS has not put silicon-based MEMS out of business. Silicon MEMS like airbag deployers and micromirror displays are sold by the millions; PDMS microdevices are not.

Maybe PDMS did not quite become the future of microfluidics, but it did help shape it. As a consolation prize, it transformed how we think about microfluidic automation and microdevice prototyping in our labs—two features that would turn out to be essential to the development of organs-on-chips (chapter 6).

3D-PRINTED MICROFLUIDICS

Even though PDMS freed many researchers from the aristocratic regime that forced them to use centralized facilities for each device, scientists still had to use a facility to fabricate a mold every time they wanted to make a new prototype or improve its design. It takes some time to implement a mature democracy. The advent of *3D printing*, a set of techniques collectively known by engineers as *additive manufacturing*, would truly allow microfluidic engineers not only to make the mold outside of a clean room, but also print a complex, impossible-to-mold 3D channel design from a digital file—and share the design online. Perhaps just as important, 3D printers automate the fabrication process, making them a cost-effective and precise replacement to previous manual methods. With 3D printing, you do not need to be a microfluidic engineer to fabricate a microfluidic device—you only need to have access to the same printer (or similar) as the microfluidic engineer who designed it.

The realization that microfluidic devices could be 3D printed—as evident as it seems now—did not come all at once to the minds of microfluidic engineers. Microfluidic devices used to be as flat as a painter's canvas. The need to 3D print them came initially from the need to make them more three-dimensional. Philippe Renaud was the first micro-sculptor that made it possible for engineers to lift the fluids and walls out of the plane of that original canvas. His group started building 3D microfluidic devices by lamination of photopatterned layers. Then he started using 3D printers for microfluidics. In fact, people had not coined the name 3D printer yet—back then, they were known as *stereolithography apparatuses* (*SLAs*).

SLA 3D printing is a layer-by-layer printing process invented by Japanese researcher Hideo Kodama in the early 1970s that uses UV light to cure photosensitive polymers (called *resins*). A typical resin consists of a photosensitive monomer and a *photoinitiator* molecule. The photoinitiator initiates the polymerization reaction when the printer shines light of a specific wavelength (typically UV). Three French scientists, Jean-Claude André, Olivier De Witte, and Alain Le Méhauté, submitted a patent for a stereolithographic process on July 16, 1984, but their company decided to abandon the patent.

Just a few days later, on August 1, 1984, Chuck Hall submitted a similar patent where he coined the term *stereolithography*. 3D Systems, the company founded by Hull to commercialize SLA 3D printing,[70] pursued the patent. 3D Systems launched the SLA-250, their first SLA printer, in 1988. However, its cost was not accessible to single labs. Microfluidic engineers would have to wait for the broader availability of SLA printers—either in the form of commercial services or affordable desktop SLA 3D printers—and to develop biocompatible resins. The first pioneers, like Philippe Renaud, could not wait and developed their own printers (box 5.7).

Renaud entered the microfabrication field somewhat fortuitously. In 1993, he got an offer to join the *Centre Suïsse d'Électronique et de Microtechnique* (*CSEM*) in the sensors team. "I was in charge of developing accelerometers despite having never seen a wafer before," Renaud half-jokes.

BOX 5.7: BIOGRAPHY
Philippe Renaud

Philippe Renaud's ancestors have been winemakers for centuries. He was born in 1958 in a small town surrounded by vineyards named Cortaillod near Neuchâtel in western Switzerland. Cortaillod was a Neolithic and Bronze Age settlement as well as a medieval village before growing into a town, so it has a lot of history. "When you work with nature and live with what you grow, you stay humble because there are good years and bad years," Renaud says as he ponders his past. Growing up, Renaud always thought he would go back to the vineyards after his high school studies, but it soon became apparent that, with mechanization, they would need more land to stay competitive. During high school, Renaud was fascinated by physics, so he decided to study theoretical physics at the University of Neuchâtel, next to Lake Neuchâtel. For his PhD, also in theoretical physics, Renaud moved in 1983 to the University of Lausanne, located one hour south by the shores of Lake Geneva.

Something urged Philippe Renaud to leave theoretical physics and his beloved lake-spotted corner of Switzerland in 1984. That year, Heinrich Rohrer, who was at IBM Zurich, was invited to give a seminar at Lausanne. In his seminar, Rohrer showed something that made a deep impression on PhD student Renaud: the first atomic-resolution images of silicon, nickel, and gold surfaces with a scanning tunneling microscope (STM) (see box 4.13). As an undergrad in physics, "I had been told you could not see atoms, but here was this man showing atoms with a simple machine," Renaud vividly recalls.

After he finished his PhD in 1988, Renaud joined John Clarke's STM group at UC Berkeley for a postdoc, although he was not away from Switzerland for long. He obtained a position in Rohrer's group at IBM Zurich in 1990, sharing the lab with Christoph Gerber. Renaud evokes it was an exciting time to be at IBM Zurich. Gerd Binnig and Heinrich Rohrer had received the Nobel Prize in Physics for the invention of the STM in 1986. The following year, Johannes Bednorz and Alex Müller, also from IBM Zurich, had received the Nobel Prize in Physics for discovering high-temperature superconductivity in ceramic materials. "IBM [Zurich] is very small. There is one cafeteria, you meet everybody there. You have four people holding two Nobel prizes in front of you, and you are drinking coffee with them," Renaud remembers fondly. He learned to build instruments because STMs were not commercially available back then. "I enjoyed it very much. Someone told me: this is engineering," Renaud remarks with emphasis, making sure I get the irony.

In 1993 he became assistant professor in microengineering at EPFL. Renaud decided to start working on packaging and nonplanar devices—"the third dimension," as he calls it. Chemist Arnaud Bertsch was just finishing his PhD thesis on micro-stereolithography at the University of Lorraine in Nancy (France) with Jean-Claude André—the inventor whose patent had been abandoned. Bertsch joined the group in early 1997. His setup at Nancy was based on scanning a laser beam over the surface of a polymer vat. "It would be good to flash the entire surface with a micromirror display," Renaud proposed to Bertsch. Renaud was aware of the MEMS-based digital micromirror displays (DMDs) developed by Dr. Larry Hornbeck's team at Texas Instruments. Hornbeck had worked on the DMD since 1977, but Texas Instruments started commercializing it only in 1992. At the time, the DMD was sold as an isolated component—projectors were not available. Renaud and Bertsch acquired a VGA 800 pixel × 600 pixel display prototype and built a UV projector that could adjust the size of the image. To this day, this size adjustability remains unique. Commercially available desktop 3D printers now incorporate the Bertsch–Renaud projection setup based on the Texas Instruments projector, but unfortunately downgraded to fixed image sizes.

One of the first prints was a tiny Statue of Liberty. Due to a rendering mistake, the printed micro-statue came out of the bath holding the torch with her left hand, but otherwise the detail was excellent. The system was published in 2000, demonstrating a resolution of 3 microns in *X*, *Y*, and *Z*.[71] For the lecture for Renaud's promotion to professor, in 1997, "we printed a red wine glass with one nanoliter capacity at 3-micron resolution, and I told the attendees: 'you can have up to three glasses,'" Renaud explains with the same wisecracking tone that he must have used that day. In 2000, Harp Minhas announced his plans to start the journal *Lab on a Chip* and asked Renaud whether he would submit a paper for his new journal. Renaud turned to Bertsch for suggestions on what they could print that would suit *Lab on a Chip*. "We could print a [3D] mixer," proposed Bertsch, who knew that microfluidic mixing of laminar flows was a hot topic. They used a homemade, transparent acrylate resin mixed with a high-absorption photoinitiator to 3D print a 3D mixer that achieved mixing of two streams in the space of ~3 millimeters. The first 3D-printed microfluidic device appeared in the first issue of *Lab on a Chip* in 2001 (figure 5.13).[72]

Renaud says he never regarded 3D printing as a manufacturing method. "I saw it as a prototyping technology," he says. "My concern was that if you do not have a way to manufacture something, it will be useless to prototype it. That's why we got interested in laminates," he adds. At some point, they had a project with the watch industry that consisted of miniaturizing gears using the LIGA technique (which uses synchrotron radiation). They did not have access to a synchrotron, which forced Renaud to think about an alternative—what he calls "a poor man's LIGA." They started to work with dry-film resists and polyimide used for microelectronics, but that was not enough for the watchmakers because they demanded perfectly vertical sidewalls. "We noticed that IBM had developed an epoxy for MEMS applications," he explains. Renaud refers to SU-8, which produces vertical sidewalls (see chapter 5). They published a few papers on "photoplastic microfluidics" (in Renaud's own words) in 1997 and 1998 with SU-8 laminates.[73] Every microfluidics textbook now covers Renaud's lamination strategy.

With time, 3D printing and lamination—both pioneered by the same Swiss researcher born in a vineyard—have become two complementary and widely used strategies in the fabrication of complex microfluidic systems such as organs-on-chips. Please do raise your next glass to Renaud, to honor not only his contributions but also his playfulness and winemaking passion: he went as far as to design and fabricate a microfluidic chip that continuously makes wine from grape juice![74]

3D PRINTING: A TOOL FOR BEING CREATIVE WITH THE 3D DESIGNS AND THE MATERIALS CHEMISTRY

3D printing can also be a tool that stimulates the creativity of students. In 2012, my student Anthony Au was going through a rough time in his PhD

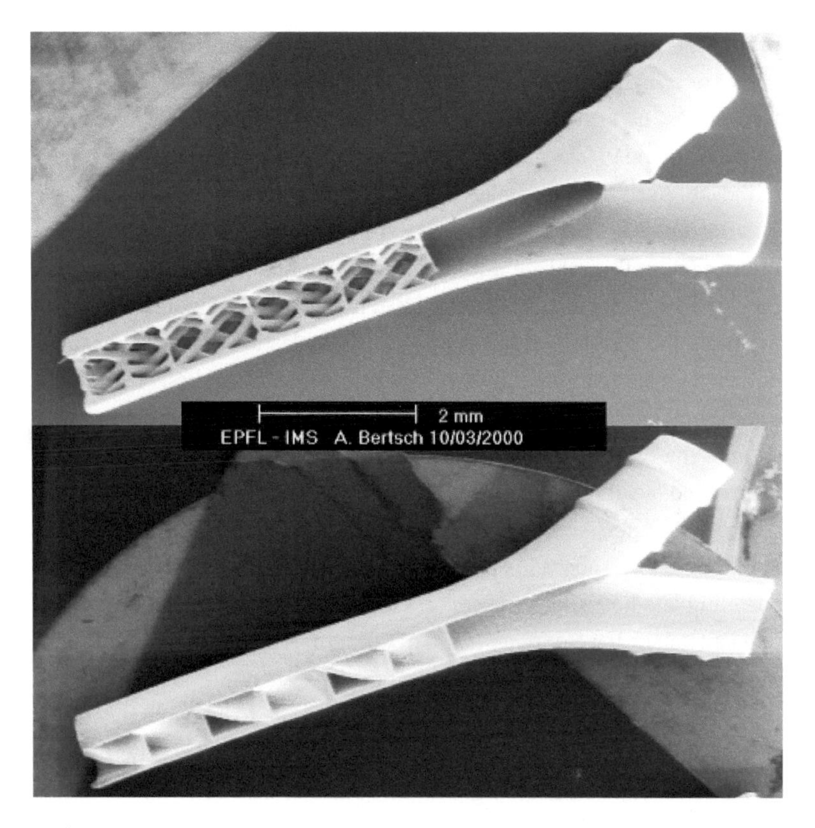

Figure 5.13
The first 3D-printed microfluidic devices were these 3D microfluidic mixers printed by Arnaud Bertsch and Philippe Renaud using a stereolithographic 3D printer. The devices have two inlets (top right) and one outlet (bottom left). The inside of the devices contain 3D features that enhance mixing. The walls of the devices are made of a transparent resin that is opaque to electrons; hence, they appear as white in these scanning electron micrographs.
Images courtesy of Philippe Renaud and Arnaud Bertsch.

project, mostly because I had given him an unreasonably challenging task. We were trying to build automated cell culture chambers in PDMS that would allow us to switch the perfusion of drugs using PDMS valves. The assembly of the PDMS parts was complicated and unreliable. In one corner of the lab, Anthony built a pile of the devices that did not work and it soon became ten, twenty, forty times higher than the devices that did work (see figure 5.12). At this rate, he would not finish his PhD. One day, I received a brochure from a fabrication service I had not heard of—FineLine Prototyping (now ProtoLabs)—showing a picture of a plastic-looking, transparent device, with an intriguing caption: "3D printed microfluidics." I showed it to Anthony and asked him to look into whether they would be able to print valves—that would address both the tyranny of pipetting and that of the centralized facilities.

One week later, Anthony had transformed into an entirely new student. Suddenly, excited about his project, he was running in and out of the lab, spending late hours in the evening. Unbeknown to me, he had a hidden, fantastic talent for computer-based 3D design. Anthony reminded me of those deserts that bloom overnight after they get rained on—the fertile seeds were hidden one layer beneath. He was able to generate ingenious microfluidic automation 3D modules of increasing complexity, such as valves, two-way switches (figure 5.14), pumps, and four-way switches (figure 5.15) that looked radically different from the traditionally flat microfluidic chips. And, importantly—for the first time—these microfluidic devices could be ordered by mail.[75]

Meanwhile, Adam Woolley and Greg Nordin at Brigham Young University had started concocting new resins with superior manufacturability and similar biocompatibility compared to PDMS (see box 5.8) and also worked on valving[76] and on improving the resolution—which will always improve with time because the projector's pixel density follows Moore's law. As you can imagine, we—and many others—do not use PDMS molding anymore.

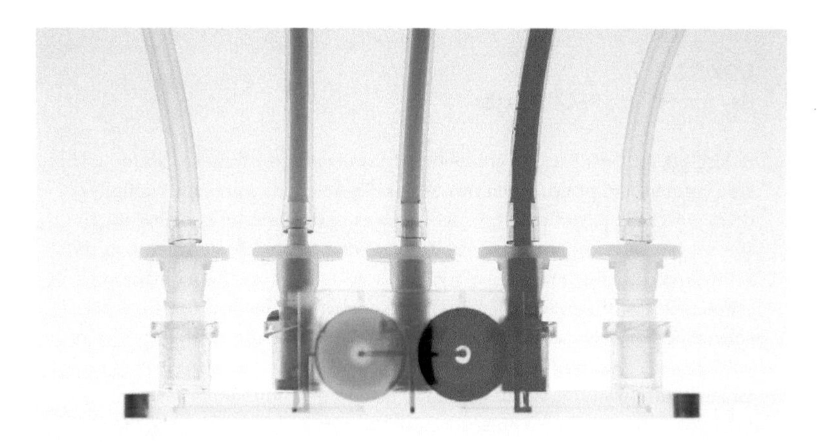

Figure 5.14
A 3D-printed, two-way pneumatic microswitch. The microswitch is composed of two 3D-printed circular microvalves.
Image courtesy of Anthony Au and Albert Folch, University of Washington.

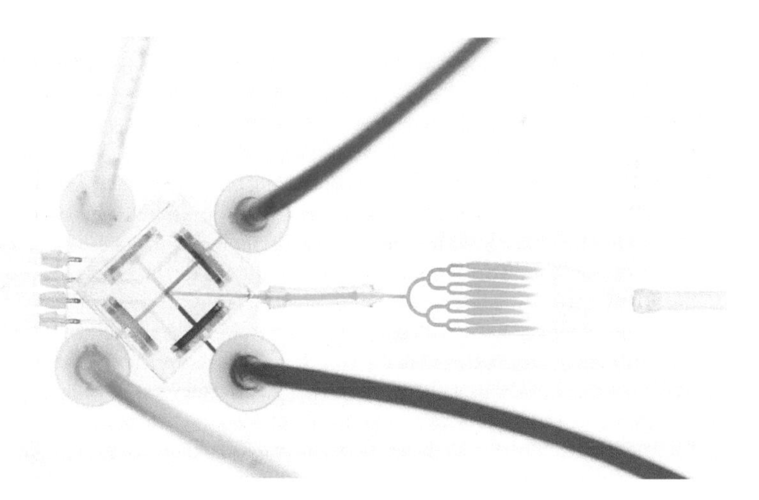

Figure 5.15
A 3D-printed, four-way pneumatic microswitch connected to a 3D-printed perfusion chamber, both pictured from above. The microswitch is composed of four 3D-printed microvalves.
Image courtesy of Anthony Au and Albert Folch, University of Washington.

BOX 5.8
Beyond PDMS: PEG-DA-258

Before 3D printers became commercially available, researchers had to build their printers and prepare their own resins. These efforts were only worth it for fields where the payoff was very high, such as in the fields of biomaterials, tissue engineering, and drug delivery. In particular, photopolymerization, or the ability to control the formation of hydrogels using light, has found many applications. One of the most widely used photopolymerizable hydrogels has been *poly(ethylene glycol) diacrylate (PEG-DA)*. PEG-DA is a chemical derivative of poly(ethylene glycol) (PEG), a biocompatible polymer that is prescribed by doctors to clean up your intestines. The two diacrylate groups at the two ends of a PEG-DA molecule make it photosensitive. At large molecular weights, the PEG chains in PEG-DA associate with water to form a hydrogel. However, in its lowest molecular weight available, MW = 258, PEG-DA polymerizes in the absence of water to form an impermeable dry plastic. We call it *PEG-DA-258*.

Adam Woolley was one of the researchers that first became interested in PEG-DA-258 because proteins do not stick to its surface.[77] After his PhD with Rich Mathies (chapter 3) and a postdoc, Woolley was soon attracted back to microfluidics and to Brigham Young University (BYU), his alma mater. Because he was from Provo (see Box 3.2), he was returning home. "One of the greatest draws to BYU was that they had a microfabrication clean room, and unlike most universities, they didn't (and still don't) charge any access fees. That was a huge cost advantage for making microfluidic devices," recalls Woolley. Since he had worked extensively with glass microfluidic devices as a graduate student (chapter 3), "I became interested in alternatives [to PDMS] when I started as an independent faculty member," explains Woolley. He also used PDMS when he needed a flexible material, and glass when he needed something that was solvent resistant. "One of our conclusions was that the best material for microfluidics depends on your objective," says Woolley. Although PEG-DA-258 was 100 times less flexible than PDMS, Woolley demonstrated a micromolding process to build pneumatic microvalves based on PEG-DA thin membranes.[78] However, as soon as they tried molding it, back in 2011, they were deterred by the slowness of the fabrication.[79] That's when Greg Nordin, also at BYU (box 5.9), stepped in.

Nordin's foray into 3D printing for microfluidics began in fall of 2012. "I had become quite frustrated with the development and fabrication delays that seem to always attend clean room fabrication of microfluidic devices and wondered if 3D printing might offer a more attractive alternative," explains Nordin. However, 3D printers were not readily available, so in winter 2013 he offered a group of senior undergraduates to build a laser-based stereolithographic 3D printer from scratch. Overall, the project was a failure in that the unfavorable trade-offs between resolution, printing speed, and print area became obvious. "It was this experience that convinced me that a digital light projector (DLP)–based 3D printer offers compelling advantages over a laser-based printer."

Laser beams do not create a single dot of light—instead, they are very sharp at the center, and the intensity gradually decays away from the center. This Gaussian profile becomes wider and wider as the beam travels in space. In a laser-based 3D printer, movable mirrors steer the beam, making it difficult to minimize the distance between the last mirror and focus spot below 6–10 centimeters. "This makes it practically impossible to get micron-ish spot sizes," points out Nordin. With DLP, an entire layer is exposed at once, which provides much faster print speeds. The trade-off in the case of the DLP is that there are a fixed number of pixels in an image so one can have high resolution or large build area, but not both.

In 2014, Nordin handed off the undergraduate project to one of his graduate students. Given Woolley and Nordin's long-standing collaboration in microfluidics, they were eager to see it applied for 3D-printed microfluidics. Because the complexity of the microfluidic devices needed by Woolley was increasing, he grew frustrated with conventional microfabrication. "I had a postdoc who would spend a week in the clean room and lab creating multilayer PDMS or thermoplastic microfluidic devices, and then after a day of testing, it was back to another week of fabrication to make a few more devices," says Woolley. He realized that, at that rate, they would not be able to do all the experiments they had proposed in the grant, so he used Nordin's 3D printer instead, with all new students focusing exclusively on 3D printing. "That change ended up costing me the renewal of my grant, but Greg and I were able to write a new proposal [on 3D printing], and that was funded instead," recalls Woolley. "Nearly all our microfluidics are 3D printed now," adds Woolley. "With a commercial printer, we demonstrated [in 2015] the first 3D printed pneumatic membrane valves," proudly says Nordin.[80]

* * *

3D printing obviously offers a lot of 3D fabrication freedom to microfluidic engineers, but, perhaps as important, it is organically applicable to "cloud manufacturing,"[81] a means of production where fabrication of products is distributed over a network of small-scale, decentralized nodes. Unlike present mass manufacturing based on rigid supply chains, cloud manufacturing is based on agile small manufacturers that could respond

BOX 5.9: BIOGRAPHY
Greg Nordin

Greg Nordin and Adam Woolley form a unique research partnership. At a professional level, they share a passion for 3D printing; outside the lab, they share their Mormon faith and Mormon family backgrounds—an important factor in their ending up at BYU, the Mormon-run university in Provo, Utah. Nordin was born in Mesa, Arizona, and grew up in San Jose, California. He became interested in physics in high school, and, having grown up in a Mormon family, BYU was a natural choice for him to study physics. "I had a great undergraduate experience at BYU, the best part of which was meeting and marrying my wife!" recalls Nordin. He graduated from BYU in 1984 and started his graduate studies in cosmology and astrophysics at the University of California, Los Angeles (UCLA), but he was interested in optics and photonics. He switched to the University of Southern California across town—where they had an outstanding program—after completing his masters at UCLA. To complete his degree and provide for his family, Nordin worked part time for Hughes Aircraft Co. (with fellowships from Hughes) until he obtained his PhD in 1992. "It was costly in terms of time, but it was the best academic decision I ever made," he reckons. Nordin left Hughes and joined the electrical engineering department at the University of Alabama in Huntsville (UAH). He started building microfluidic devices in 2002–2003 for a local company and has been doing microfluidics ever since. Nordin was eventually attracted back to BYU for family reasons. "My wife and I had very fond memories of our time at BYU, so it was delightful to be back as a faculty member," remembers Nordin.

swiftly to shifting inventories and market demands.[82] Recent work by a team led by Rashid Bashir and Brian Cunningham at the University of Illinois at Urbana-Champaign, offers a hint of this possible future for microfluidics. They used a stereolithographically 3D-printed microfluidic cartridge (figure 5.16) for the detection of COVID-19 viruses from a nasal swab and developed a smartphone-based reader that gives the results in thirty minutes. The researchers designed the channels as open grooves and manually closed them with biocompatible transparent tape. The 3D-printed microfluidic device performs amplification of nucleic acids by a method called reverse transcription loop-mediated isothermal

Figure 5.16
A 3D-printed microfluidic device for detection of COVID-19 virus from nasal swabs. The circle in the device is around 1 centimeter wide.
Image courtesy of William King, Rashid Bashir, and Brian Cunningham, University of Illinois at Urbana-Champaign.

amplification (RT-LAMP) (see box 3.3).[83] LAMP is essentially the constant-temperature version of PCR, the nucleic acid amplification method by thermal cycling. LAMP works at 65°C and is faster and cheaper than PCR. The researchers only had to warm up the 3D-printed cartridge to 65°C for thirty minutes to amplify the COVID-19 nucleic acids in the nasal swabs. The device bypasses the need for sending samples to a centralized lab. "We could test for COVID-19 at public events, auditoriums, large gatherings and potentially even at home for self-testing," Bashir says.[84] And of course, the design can be freely shared via the web or email for others to print copies. Continuing the democratization movement that was triggered by PDMS, 3D printing has finally wiped out all the elitist restrictions of microfluidics and promises to deliver a new, blossoming microfluidics era (box 5.10).

BOX 5.10
Multi-material 3D Printing Techniques

The early multi-material shortcomings of stereolithography spurred the development of alternative, multi-material 3D printing techniques. One such technique is *fused deposition modeling* (*FDM*) or thermoplastic extrusion, a form of 3D printing invented by Scott Crump in 1988. FDM is based on extruding a heated thermoplastic material from a motor-driven nozzle head that can move in three dimensions. The material hardens by spontaneous cooling immediately after extrusion. FDM printers are mechanically simple and thus relatively inexpensive and user-friendly, around which an open-source community of designers has flourished (e.g., the Thingiverse website[85]). While resolution in FDM is still a challenge, the technique's biocompatibility and multi-material versatility are unparalleled, a feature that has attracted the attention of many microfluidic engineers (figure 5.17). FDM is sold in convenient filament rolls of cheap and biocompatible thermoplastic polymers, such as acrylonitrile butadiene styrene (ABS), polylactic acid (PLA), polycarbonate, polyethylene terephthalate glycol-modified (PETG), thermoplastic elastomers (TPEs), thermoplastic polyurethane (TPU), poly(methyl methacrylate) (PMMA), polypropylene, nylon, polyamide, and polystyrene. Plastics that are biodegradable, water-soluble, conductive, ferromagnetic, metal-colored, glow-in-the-dark, thermochromic, and ceramic are also available.

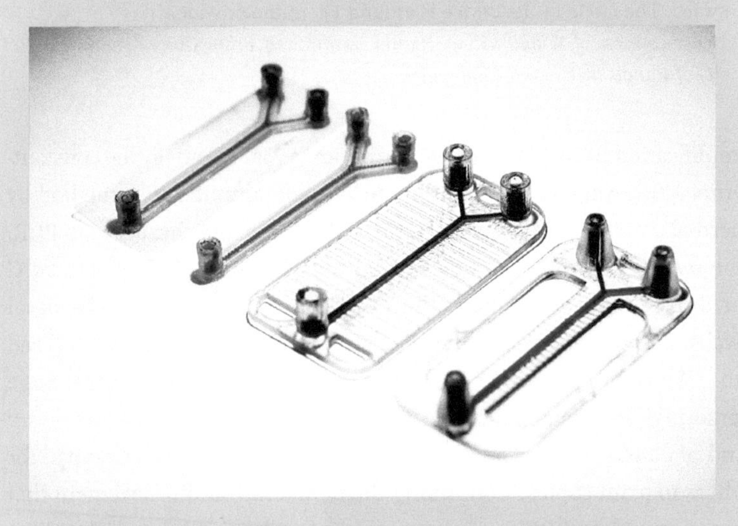

Figure 5.17
Microfluidic devices printed by FDM by Niall Macdonald and Michael Breadmore.
Image courtesy of CBMS (Okaar Photography).

A strategy related to FDM is called *direct ink writing* (*DIW*). In DIW, the nozzle extrudes liquid precursors at room temperature such as metallic solutions, hydrogels, and cell-laden solutions through a nozzle head.

Microfluidic devices created by FDM or DIW are appealing in principle because one can build them in many materials with the simplicity of pressing a button. However, microchannel fabrication with both these techniques has lacked the necessary resolution so far. First, the size of the extruded filaments is larger than typical channels used in microfluidics, although "millifluidic reactionware" is possible.[86] The lack of structural integrity between the layers can result in weak seals.[87] Also, from a topological consideration, it might not be possible to lay down the walls of any arbitrary channel layout in the form of extruded filaments, especially at channel intersections where joining filament ends might cause leaks. Still, very simple layouts, such as for electrochemical detection,[88] have been demonstrated. Notably, Michael Breadmore's group at the University of Tasmania used a dual-extruder FDM printer to build an integrated 400-micron-wide porous barrier (using a composite Lay-Felt filament) separating 700-micron channels made with ABS.[89] The composite Lay-Felt barrier turned porous after the dissolution of the water-soluble component within the barrier. And Bastian Rapp's group at the University of Freiburg recently used PMMA filament to 3D print microfluidic chips on PMMA surfaces, achieving an approximately 300-micron channel resolution by using a commercially available, inexpensive FDM 3D printer.[90] The chips by both groups are remarkably transparent, so the technique has potential for applications that, requiring less resolution, benefit from the printing of multiple materials.

6 ORGANS-ON-CHIPS

Dan Huh, a postdoc at the Wyss Institute at Harvard, had been thinking very hard for a few months about the design of a microfluidic device that mimicked the biomechanical functions of the lung. Ironically, the inspiration for a critical feature—a flexible flap that allowed him to pneumatically inflate the device and replicate lung inhalation—came to him while dozing off at church.

This lung-on-a-chip device belongs to a major class of microfluidic devices, called *organs-on-chips*, that attempt to mimic the physiological and disease mechanisms of human organs, including tumors. By directly testing drugs in small pieces of human tissue, scientists hope to obtain relevant information about diseases and drug efficacy in humans, without having to test the drugs first in animals. Smallness is important. We will not be testing drugs in whole organs bubbling in a glass jar any time soon as Hollywood would like you to believe. A large organ must be properly vascularized and perfused to be kept alive and functioning, whereas a small piece of tissue can be maintained with the help of microfluidics.

Clever design is key. The 3D structure and functional capabilities of Huh's device went so far as to impress Paola Antonelli, a curator for the

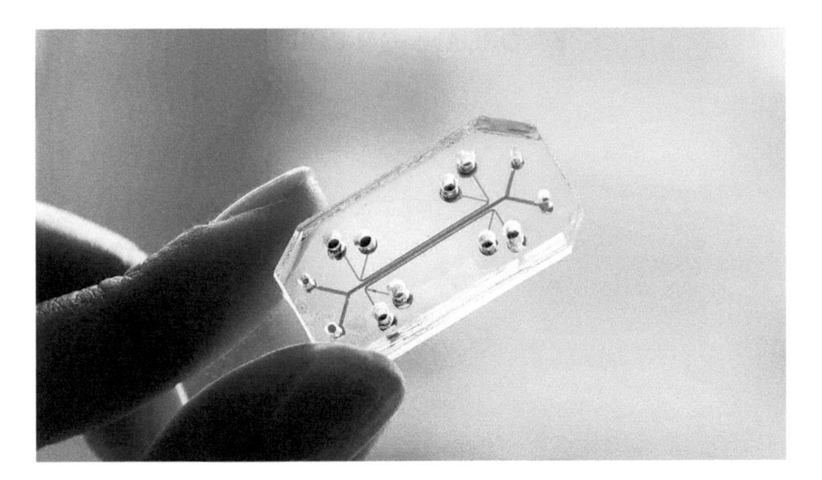

Figure 6.1
This lung-on-a-chip device is composed of three PDMS layers and has won several design awards.
Image courtesy of Wyss Institute, Harvard University.

Museum of Modern Art (MoMA) of New York City. She formally acquired a chip replica that was exhibited at MoMA for one year during 2015 (figure 6.1). That same year, the Wyss Institute's human organs-on-chips, represented by the human lung, gut, and liver chips, won the Design of the Year 2015 Award, the United Kingdom's most prestigious design award.

It took many years of developments by scientists and engineers to learn to build organs-on-chips. Before attempting to mimic tissues, researchers had to make sure that they could keep cells alive in microchannels. The materials used for the inkjet nozzle and the DNA analyzers of the Human Genome Project—glass and silicon—were not permeable to oxygen, so cells did not survive long inside a glass or silicon channel. Researchers found the right material in PDMS (see chapter 5).

THE ORGAN-ON-A-CHIP REVOLUTION

If you have been severely sick, you will have felt the unfairness of the high cost of drugs. I'm not talking about generic drugs like aspirin or valium, which have been around for decades, but the latest drugs for cancer, cardiovascular or gastrointestinal diseases, nervous disorders, and rare conditions. These drugs are so costly that even selling an expensive car will not pay for a year's supply of them.

The question before us, then, is whether we can apply technology to make drug development drastically less expensive. The most poignant example is that of cancer drugs, whose approval is by far the most wasteful process of all drug types: the FDA only approves less than 4 percent of cancer drugs.[1] That means that 96 percent of cancer drugs spend more than a decade being tested in petri dishes, mice, and a small set of patients before scientists finally realize that they are *not* suitable for human use. Each drug in the 4 percent that gets approved bears an average price tag of more than a billion dollars, largely because of the wasteful 96 percent.[2] If one could come up with a technology that gets rid of much of the 96 percent, the cost of drug development would be reduced by as much as twenty-fold. Pharma would likely save additional money because resources could be concentrated on the remaining 4 percent to accelerate the drug development process. When we consider all drugs (both cancer and non-cancer drugs), about 86 percent of the research and development spent on drugs is wasted.[3] The good news is that this is where microfluidics can help.

The fundamental problem is that, for the last century, biologists have been performing preclinical drug tests in petri dishes and animals for lack of better guinea pigs. But you can immediately see that you do not look like a mouse. The mouse heart weighs only about 0.2 grams and has a heartbeat of around 500–600 beats per minute, whereas the human heart weighs around 250–300 grams and beats, on average, 60–70 times per minute. Like in humans, mice have five lobes in their right lung, but unlike humans, mice only have a single left lung. And differences between species in metabolizing enzymes in the liver lead to different drug toxicity profiles in rodents versus humans.[4]

The old tenet for testing drugs in mice has been that mice and humans are genetically similar, which is disturbingly accurate for the most part.

However, it is less accurate for cancer than for other diseases: our tumors arise from mutations that evolve in our organism and grow differently for each individual into a unique, aberrant microenvironment that was not encoded in our original DNA. It is not surprising that mouse tumors often do not respond to drugs the same way the equivalent human tumors respond. Furthermore, drug toxicity and dosing differ greatly between human and animal models. The official conclusion is that the preclinical animal tests required by the FDA do not accurately predict toxic doses and drug metabolism observed in humans.[5]

Why do pharmaceutical companies place so much hope that a cancer drug designed to work in mice will also work in humans if the odds are twenty to one against them? The answer is because they do not have better testing tools, and they pass the bill down to the patient-customer to recover the investment. Researchers sacrifice millions of animals each year in such tests, a practice that has sparked fierce and growing opposition from animal-rights groups. The good news is that pharma agrees that this testing paradigm is immensely wasteful and that we need new testing systems that are physiologically closer to a human.

One approach has been to improve the petri dish by enticing the cells to grow in 3D spheres or spheroids (sometimes also termed organoids because the cells behave as if they were part of mini-organs both in sickness and in health). Because spheroids are so small—typically 100–200 microns in diameter—it is possible to generate a lot of spheroids with small patient biopsies and place them in microfluidic arrays for multiple drug treatments (figure 6.2).[6] However, organoids are randomly organized and are too small to be part of a circulatory system that provides their access to nutrients in a stable manner—such as that seen in tissues, which are layered with blood vessels that connect to the rest of the body.

Another approach—the organs-on-chips—builds on the organoids concept with microfluidic devices that contain human cells in 3D architectures to better recapitulate the health and disease processes of human organs outside of the human organism. Organs-on-chips are the perfect guinea pigs—they are inexpensive to maintain via microfluidic perfusion, they recreate tissue interfaces that are therapeutically relevant (such as

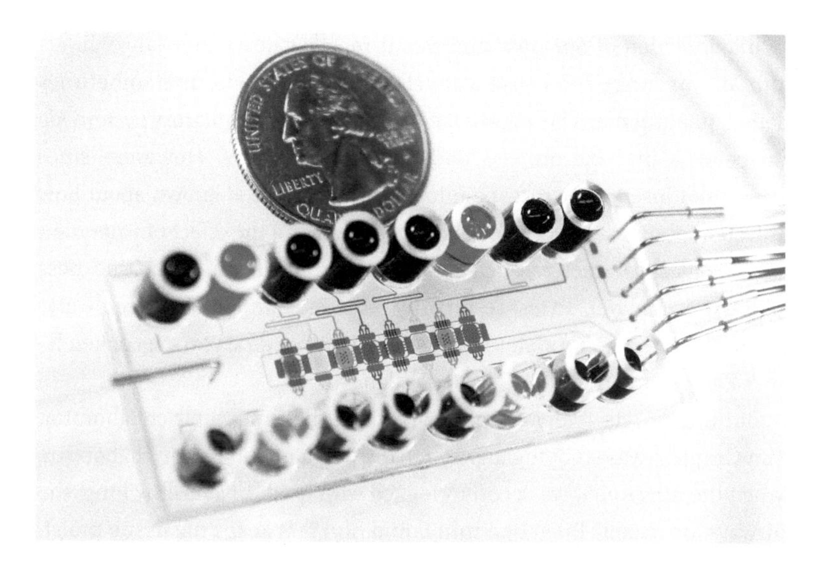

Figure 6.2
A microfluidic platform for cultivating ovarian cancer spheroids and testing their responses to chemotherapies.
Image courtesy of Alan González-Suárez and Alex Revzin, Mayo Clinic.

the blood–brain barrier or the capillary interface in the lung), and they can model new modes of drug delivery without harming a human. With these platforms, researchers have cultured gut bacteria in contact with human intestinal cells, or immune cells in contact with human lung cells, and modeled how these satellite cells play a role in normal conditions compared to various disorders. The pharmaceutical industry is paying closer and closer attention to these new devices. A successful example is the lung-on-a-chip developed by Dan Huh that could bring insight into how smoke from cigarettes damages our lungs, how viruses infect them, and whether new vaccines may work.

THE LUNG-ON-A-CHIP

The lung is a complex organ that acts as the key battleground for respiratory pathogens such as the COVID-19 virus trying to enter our body.

While infection of our nose and throat mostly causes annoyance, infection of our lungs can cause a much more dangerous, and sometimes lethal, pneumonia. The virus appears to enter our respiratory system via suspended, invisible droplets and infects our airways. However—since we cannot insert a microscope into our lungs—little is known about how infection occurs at the cellular and tissue level and the effect of important physiological parameters, such as the speed of inhalation or the thickness of the mucosa layer. To test respiratory diseases, Dan Huh has been working on a *lung-on-a-chip* system that he developed several years ago when he was a grad student in Shu Takayama's lab.

But the lung is a very complex organ. At the time, their collaborator, Jim Gropper, was studying airway narrowing, a phenomenon that occurs when the lung's airways become clogged with fluid. In a healthy lung, the airways are usually lined by a thin liquid film that keeps the tissue moist. Under certain conditions—for example, due to inflammation during the attack of a specific pathogen like a virus—the tissue reacts by producing much more fluid, which causes the airways to narrow. When that happens, there is a higher chance of the airway being occluded by a liquid plug that moves down the airways until it ruptures. Doctors like Gropper can hear the plugs rupturing as a crackling sound with a stethoscope and infer how your lungs are doing by listening to them.

Gropper had been studying the fluid mechanics of crackling, and, as soon as Huh came back from Stanford (box 6.1) in 2004, Gropper thought it would be compelling to have a microfluidic model of crackling. The first step was to see if they could culture primary airway cells in a microfluidic device and keep them alive. Airway cells, as the name suggests, are the cells that are in contact with air inside the lungs. Fortunately, human airway cells were commercially available.

For the design, in 2005, Huh found inspiration in Transwell inserts, a simple device that consists of a porous membrane where cells are seeded. Researchers had previously shown that lung epithelial cells (the airway cells) could stay alive grown on the porous membrane of the inserts while exposed to air from the top and fed with nutrients dissolved in cell

BOX 6.1: BIOGRAPHY
Dan Huh

Growing up, Dan Huh always liked the idea of building things. He was born in 1976 in Jeju, a small subtropical island of volcanic origin south of South Korea. For college, he went north in 1994 to the mainland to study mechanical engineering at Seoul National University. There, he learned about thermodynamics, heat transfer, car engines, and so on, "but towards the end of my undergraduate studies, I found I was more attracted to living things," he remembers. During his senior year, Huh worked in a lab that studied turbulent flow through heart valves, which got him interested in biofluid mechanics. He also worked in the lab of Dong-Chul Han on how fluid buildup and drainage failure lead to glaucoma in the eye. Although he left soon afterward, his interest in ocular research would stay with him for many years. It was the year 2000, and biomedical engineering was becoming popular in Korea and around the world. He decided to study abroad for his PhD, and he was accepted into the University of Michigan.

At Michigan, Huh met Shuichi Takayama, a young assistant professor who had just finished his postdoc in the Whitesides lab. He started working in the Takayama lab in various projects. One included a collaboration with Jim Gropper, an MD/PhD expert in biofluid mechanics that was investigating the respiratory system. After a year and a half, however, Huh walked into Takayama's office and gave him the bad news: he had decided to apply to the Stanford mechanical engineering program to be with his girlfriend, who was at Stanford at the time.

Takayama respected Huh's personal decision, and Huh got an MS degree in bioengineering from Michigan before leaving for Stanford in 2003. Once at Stanford, however, things did not work out as expected. Huh and his girlfriend eventually broke up. Most important, although he was doing some microfluidics research, "I think I really missed working with Shu [Takayama], his creativity, his support, and the great environment he had created for us at Michigan," says Huh. After another year and a half, Huh decided to move back to Michigan in 2004 to complete his PhD. "Shu and Jim Gropper were glad to have me back and got serious about the idea of developing a microfluidic model of the lung," he recalls. Back at Michigan, he developed the first lung-on-a-chip.

culture medium from below. To imitate the Transwell structure, he put a PDMS channel on top, another one at the bottom, and sandwiched a porous membrane in the middle. "That design was key to differentiating cells in a device because we had to culture them under air while feeding them from the bottom," says Huh. "It worked beautifully," he remembers. Dozens of groups have used this sandwich design since then.

The next critical question was whether plugs could be generated in a traveling fashion and whether they would imitate the crackling. Huh had worked with an air–fluid interface for a previous paper. In 2006, he used that technique to dynamically switch plugs on and off and to send those plugs over the cells. They found that the plugs did emulate the crackling. Furthermore, when the plugs ruptured, cells would die, "so we were able to claim that crackling might be associated with mechanical lung injury," he adds. The paper was published in 2007 in *PNAS*.[7]

Huh wanted to delve deeper into biology, so he applied for a postdoc to Don Ingber's lab at Harvard, who was very familiar with his *PNAS* paper and had become drawn to the idea of a lung-on-a-chip. Ingber had co-mentored Shu Takayama while he was in Whitesides's group and had heard a lecture by Takayama around 2005 presenting the microfabricated liquid plug generator. "I was amazed by [Takayama's] presentation because the plugs moving through the channels generated a sound that was exactly the same as the 'crackle' sound I was taught to listen for through a stethoscope during a lung exam when I was a medical student," says Ingber. "This truly was amazing to me because every med student in the world would ask their professor what caused that sound, and no professor could ever provide a good answer beyond 'fluid in the lungs,'" he adds. During Huh's postdoc interview in 2006, Ingber explained to Huh that "what would be really exciting would be to build a real living breathing lung lined by living cells with human tissue–tissue interfaces that could be visualized and studied in vitro." Huh accepted Ingber's offer to join his group.

Initially, Ingber asked Huh whether he could build a lung model to probe environmental toxicity because he had funding from a collaborative environmental project with George Whitesides. Huh did not want to build a replica of the device he had built in the Takayama lab, so he focused instead on a model of the alveoli, the small sacs at the tip of the airways where gas exchange between air and blood capillaries occurs. "That's where a lot of important things happen," insists Huh. The first step was to see if they could culture alveolar epithelial cells in a PDMS device—which worked. But the real challenge was how to imitate the dynamic contraction and expansion of air sacs physiologically. "I tried really hard to design some kind of a mechanical stretcher that could be integrated into a microfluidic device where I could culture alveolar cells and then stretch and relax them to mimic breathing motions," remembers Huh. But none of the designs he could think of would work. Until one Sunday morning, while dozing off at the service of his Protestant church in Brookline, the inspiration for the design came to him: two side PDMS chambers could be used as pneumatic actuators to pull on the porous membrane seeded with epithelial cells.

Huh presented the design at the joint Whitesides–Ingber group meeting, where it received general praise from the senior scientists. The design required the chemical etching of PDMS for one of the steps, but thankfully the Takayama group had published a paper on that rare technique. Yet the design (figure 6.3) did not capture the full picture. Ingber wanted to mimic the alveolar-capillary unit where alveolar lung cells touch endothelial blood capillary cells. In 2008, when Dan Huh showed that he could fabricate the porous membrane by molding PDMS holes, Don Ingber immediately reiterated to him the vision he had expressed during his postdoc interview two years earlier and on which others in his group had made some headway: "Can you culture vascular cells at the bottom [underneath the alveolar cells]?" Ingber wanted to use the lung-on-a-chip device to screen the toxicity of nanoparticles. For the paper, they demonstrated the toxicity of silica nanoparticles, showing that the toxicity went up when they stretched the tissue because the cellular uptake of the nanoparticles increased as well. A critical demonstration was the recruitment of immune cells mimicking the natural lung response: they placed

Figure 6.3
Lung-on-a-chip device designed by Dan Huh.
Image courtesy of the Wyss Institute, Harvard University.

bacterial cells on the channel that contained lung cells, then they introduced white blood cells into the channel that contained the endothelial cells, and showed the entire process of immune cell adhesion, diapedesis (the movement of white blood cells through capillary walls), and phagocytosis (the engulfing of bacteria by a white blood cell) in the lung compartment; these results were published in *Science*.[8] "We were able to claim that, unlike other models that contained single cells, by having multiple cell types interacting with each other, we can start mimicking more complex responses that happen at the organ level—not just at the tissue level," explains Huh.

Using the same lung-on-a-chip device, Huh next modeled pulmonary edema processes induced by drug toxicity observed in human cancer patients treated with interleukin-2 (IL-2), a type of signaling protein that regulates the activities of white blood cells. The chip allowed for reproducing mechanical forces associated with physiological breathing motions. The study revealed that these motions play a crucial role in the development of increased vascular leakage leading to pulmonary edema; circulating immune cells were not required for the development of edema. Importantly, and without performing a single animal test, this work led to identifying potential new drugs that could prevent IL-2 toxicity in the future.[9]

Dan Huh's systems caught NASA's attention. NASA wanted to discover why half of the astronauts who flew in Apollo missions reported minor bacterial or viral infections within a week of their return. Huh's lung-on-a-chip seemed ideal for testing how microgravity affects our immune system's reaction to pathogens. NASA is preparing to launch one of Huh's lung-on-a-chip platforms aboard a rocket for testing it in the International Space Station, where it will be connected to a bone marrow-on-a-chip to test its interactions with the immune system in zero gravity.

OTHER ORGANS-ON-CHIPS

The pharmaceutical industry immediately took notice as well: the organ-on-a-chip concept could be extended well beyond the lung. After Ingber and Huh published the first paper in *Science*, "we started getting inquiries and calls from pharma right away, and Don [Ingber] started a sponsored research agreement with Merck," remembers Huh proudly, since it all stemmed from his work. Don Ingber launched the Wyss Institute at Harvard in 2009 with a broader mission to develop biomimetic technologies such as organ-on-a-chip and 3D printing. After the lung-on-a-chip, Ingber had the idea of developing a gut-on-a-chip device as an extension of the lung-on-a-chip concept. During Dan Huh's second year as a postdoc in Ingber's group, Hong Jun Kim joined the lab with a background in microbiology. Kim developed the first gut-on-a-chip using the convenient human Caco-2 cell line and the PDMS stretching capabilities to mimic the peristaltic (waves of squeezing) function of the gut, a paper that raised much interest.[10]

The Wyss Institute has since generated a total of fifteen microphysiological models of living human organs, including the lung, intestine, kidney, skin, bone marrow, and blood–brain barrier, among others. To mimic the physiology of the whole human body and predict tissue responses to drugs, Wyss researchers have developed an instrument that links up to ten different organs-on-chips by transferring fluid between their common vascular channels. As a testament to its success, Wyss now employs about three hundred staff and has become the flagship for organ-on-a-chip research and biologically inspired engineering in the world. "It truly 'takes a village' to build something big," emphasizes Ingber.

Researchers around the world have followed suit, addressing various challenges. In 2010, Abe Stroock's lab achieved what seemed impossible a decade ago: developing a vascularized device made only with biological components—they used endothelial vascular cells to make the capillaries and collagen gel for the bulk of the device. For many years, the microfluidics community had placed cells in microchannels made of polymers such as PDMS. These polymers are water-impermeable, very different from the type of protein-based porous matrix where cells attach and freely move around in our body. In the body, fluid-containing walls are made of thin cell layers forming tight cell junctions. Valerie Cross and Ying Zheng in the Stroock lab molded channels in collagen gel—which is permeable to water and biomolecules—and seeded the inside with endothelial cells and smooth muscle cells until they formed a confluent layer (figure 6.4). The introduction of a dye inside the cell-lined capillary showed that the dye was fully contained inside the capillaries and did not leak into the collagen gel.[11] This finding allowed researchers to start thinking about microfluidic designs that included the vascularization and formation of tissue chambers within a chip similar to those that exist in our body. For example, a team led by Abe Lee at the University of California, Irvine (UC Irvine) has been able to construct and perfuse 3D networks of interconnected endothelial vasculatures in hydrogels within PDMS microfluidic channels (figure 6.5),[12] a device that they have used later for testing cancer drugs (see "Tumor-on-a-Chip"). Not surprisingly, flow helps form these vascular structures. Ryuji Morizane, Jennifer Lewis, and coworkers showed that kidney organoids exposed to

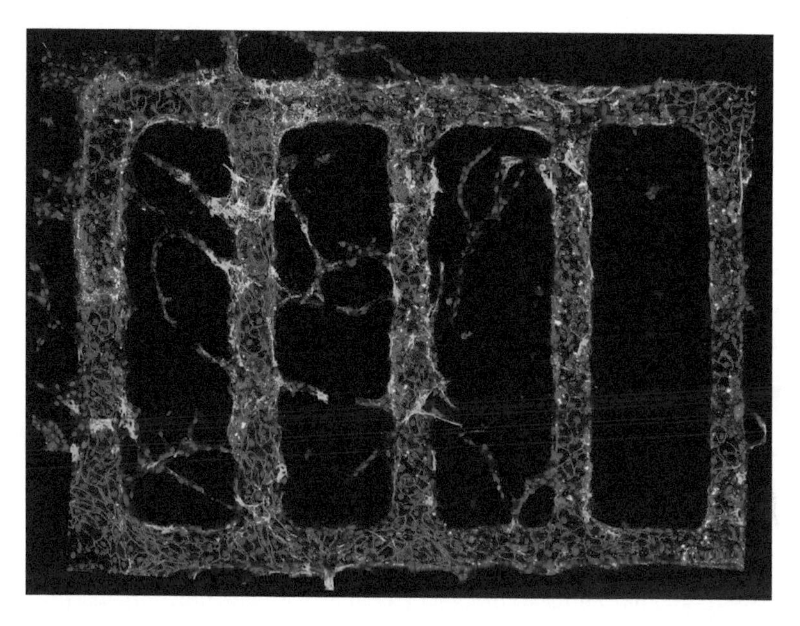

Figure 6.4
Vascularized microchannels in a collagen gel. The whole device is molded in collagen.
Image courtesy of Ying Zheng, University of Washington.

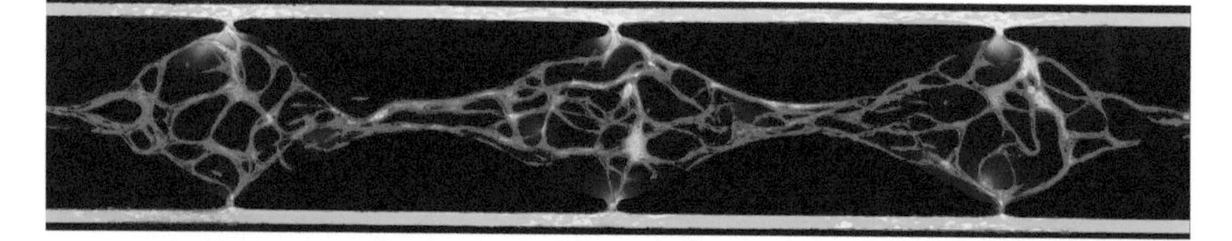

Figure 6.5
Vascularization of hydrogel-filled PDMS microchannels allows for developing drug testing platforms.
Image courtesy of Duc Phan, Xiaolin Wang, Chris Hughes, and Abe Lee, University of California, Irvine.

flow inside a microfluidic device became vascularized faster than static controls.[13]

However, it is difficult to control growth in organoids and achieve tissue homeostasis (the balance between cell elimination and cell regeneration). A team led by Hans Clevers and Matthias Lutolf demonstrated the formation of homeostatic mini-guts inside a PDMS device filled with a hydrogel collagen matrix, where a cavity in the shape of the mini-gut was laser-ablated and filled with stem cells.[14] After a few days, the mini-guts were teeming with life (figure 6.6). Inside the mini-gut, gut stem cells grew and regenerated, even after a physical or a chemical attack, reaching homeostasis. The researchers could connect the mini-intestines to a pump for perfusion, which allows for removal of dead cells and their colonization with microorganisms to model bacteria interactions essential to our gut well-being.

The liver is perhaps one of the most challenging tissues to engineer due to its 3D and biochemical complexity, connectivity with other organs, and the difficulty of keeping it in its differentiated state. Following pioneering work by Michael Shuler's[15] and Linda Griffith's[16] labs, who built early microfluidic perfusion systems for hepatic tissue engineering, in 2010 a team led by Geny Groothuis and Elisabeth Verpoorte assessed interorgan interactions in a microfluidic device using precision-cut liver slices and intestinal slices.[17] Slices preserve fundamental 3D micron-scale interactions present in the tissue architecture that cannot be reconstructed in tissue formats that are grown from cells. The researchers placed the slices in adjacent microchambers and sequentially perfused them so that the metabolites excreted by the intestinal slice were directed to the microchamber containing the liver slice. They demonstrated the interplay between the two organs by exposing the slices to bile (chenodeoxycholic acid, made by the liver). Bile exposure increased the expression of a growth factor (fibroblast growth factor 15 [FGF-15]) in the intestinal slice. FGF-15, in turn, lowered the activity of a detoxification enzyme (cytochrome p450) in the liver slice. Because one primary concern about any drug is its toxicity, pharmaceutical companies are interested in these

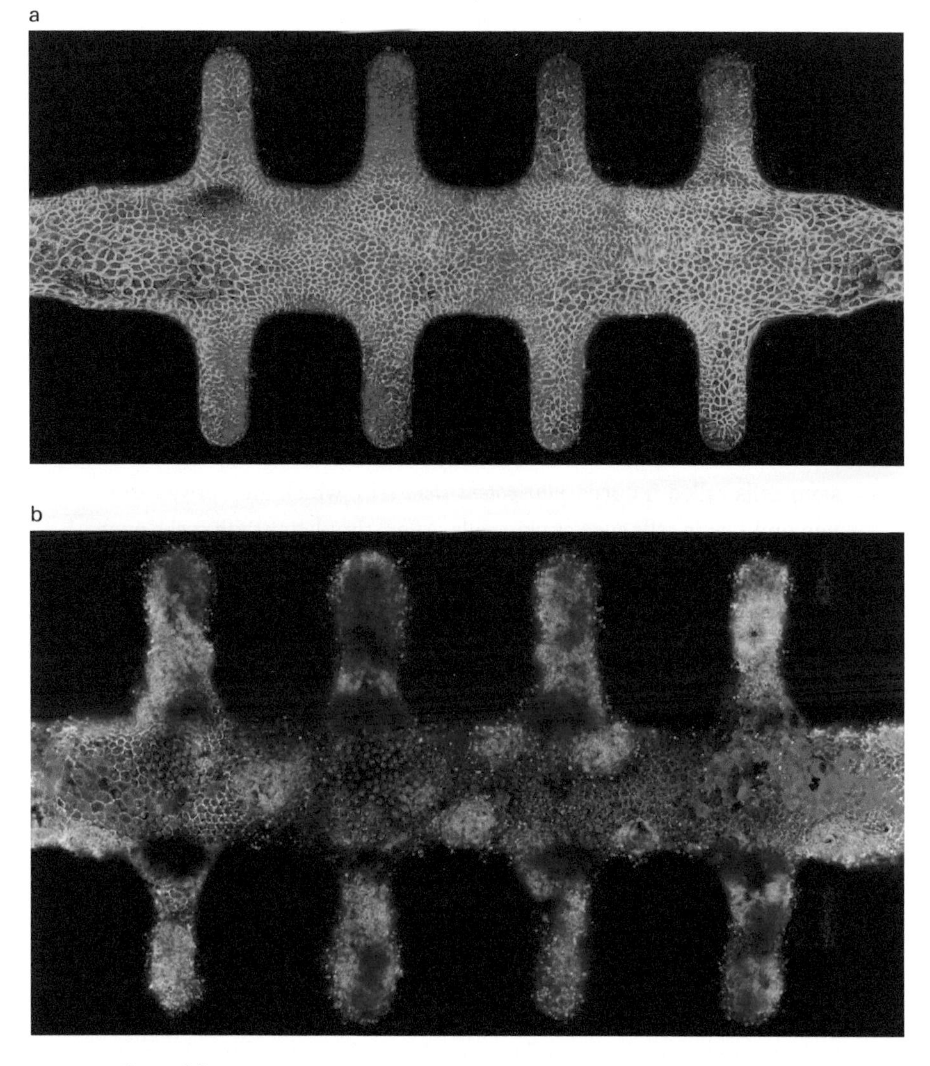

Figure 6.6a

This image of a three-day-old mini-gut shows a coherent and tightly packed epithelial sheet. Evidence of tight packing is the expression of high levels of E-cadherin at cell–cell junctions (green). The nuclei are shown in blue. The image is 1.2 millimeters wide and each arm of the device is 50 microns wide and 170 microns long.

Figure 6.6b

A fluorescence micrograph of another mini-gut after seven days in culture. The red staining shows the presence of enterocytes, a type of epithelial cells that line the inner surface of the small and large intestines, indicating that the mini-gut has formed as planned from the initial stem cells.

Images courtesy of Mike Nikolaev and Matthias Lutolf, EPFL.

liver-on-a-chip platforms that can evaluate the toxicity of compounds in human liver tissue *directly*, while potentially bypassing the costly and lengthy steps of animal testing.

Like liver-on-a-chip, heart-on-a-chip platforms containing human cells are also of vital interest to pharma because every new drug must be shown not to affect the contractility of the heart.[18] Marco Rasponi's group applied a design similar to Dan Huh's with elastomeric actuators to "push and pull" on muscle cells in a "beating-heart-on-a-chip" in 2016.[19] Since donors of human muscle tissue are rarely available, the researchers micro-engineered differentiated cardiac tissues from a type of human stem cells called induced *pluripotent stem cells*, which are derived from non-embryonic cells such as skin cells. After stimulation with cyclic push and pull, cells showed superior cardiac differentiation and promoted early spontaneous synchronous beating and better contractile capability in response to electric pacing. The researchers also demonstrated drug applications to this contractile model.

Still, organs-on-chips do not look like the organs they are trying to imitate yet. Key pieces like the microvasculature or the dynamic complexities of blood and the immune system are absent. When tissue from a human donor is available, it may be possible to bypass the complex and uncertain process of tissue microengineering and directly use a clinical biopsy. José Luis Garcia-Cordero's team from CINVESTAV-Monterrey in Mexico built a PDMS device with a large pneumatic actuator that was used to flatten a biopsy against eight open microchannels (figure 6.7). As a proof of principle, they used the device to perform a short-term drug toxicity assay on live aorta and trachea tissues from mice, but the protocol should work for human tissues as well.

To be up to date, I always keep an eye on what Dan Huh is working on. As a faculty member at the University of Pennsylvania (UPenn), Huh has rekindled his passion for ocular research by developing a blinking-eye-on-a-chip model.[20] In 2012, when applying for faculty jobs and riding the blazing success of his work at the Ingber lab, giving talks as an invited speaker at conferences and seminars, Huh received devastating

Figure 6.7

Top view of a mouse aorta biopsy (yellow rectangle, center) that has been sandwiched between a PDMS valve (below the biopsy) and a set of microchannels (above the biopsy). When the valve is actuated, the tissue is flattened against the microchannels, allowing for the selective delivery of drugs (here represented as color dyes for visualization). The tissue (2 millimeters wide and 6 millimeters long) is alive during the drug delivery procedure.

Image courtesy of Alan González-Suárez and José Luis Garcia-Cordero, CINVESTAV-Monterrey, Mexico.

news on the personal front. After battling breast cancer for ten years, his mother had been diagnosed with liver metastasis with no more treatment options. After he accepted the UPenn offer, he decided to defer the start date of his faculty job for a year and went to Korea to be with her in her last moments until she passed away at the end of 2012. "I was fortunate to spend the last month with her," he recalls. Seoul National University generously offered Huh a temporary faculty position in the biomedical engineering department while he waited to join UPenn.

During the months before Huh moved to Pennsylvania, he worked with student Jeongyun Seo to develop a model of the ocular surface. "As a mechanical engineer, I was attracted to the front of the eye because of the blinking," recalls Huh, who also remembers that they had many failures—after many trials, they built the lid in hydrogel connecting it to a miniaturized motor. They seeded corneal cells and conjunctival cells in different areas, but "culturing the cells on a rounded surface wasn't easy because the cells would roll down; using a complementary concave surface and surface tension, we were able to control the wetting of the scaffold," says Huh, taking inspiration from past challenges. Jeongyun Seo moved with Huh to UPenn, where they finally built the device that modeled the dry eye (figure 6.8). *Nature Medicine* published the study in 2019.[21]

TUMOR-ON-A-CHIP

One day in early 2000, as I was having lunch with Mehmet Toner in Boston, the conversation drifted to the relevance of animal research for human medicine because some animal-rights protesters were parading down the streets. "If we researchers had always been forbidden from using animals for research, by now we would have found a way to answer all these questions directly in humans," Toner reflected. The thought that we might be using animals out of convenience rather than out of necessity has stuck with me ever since. About a decade later, the elegance of the Wyss Institute organ-on-a-chip designs inspired a new line of inquiry in several labs, including mine, focused on cancer. Would it be possible

Figure 6.8
A blinking eye on a chip that incorporates live corneal and conjunctival cells (center yellow circle) as well as a mechanically activated hydrogel lid (blue slab) and a tear channel (blue) for periodic humidification.
Image courtesy of Dan Huh, University of Pennsylvania.

to use microfluidics to test multiple cancer drugs on a small, live tumor biopsy? That would be a more efficient, direct-in-human test for drugs and would avoid unnecessary animal suffering.

I sought the help of my cancer collaborator Ray Monnat Jr. from the neighboring pathology department at the University of Washington in Seattle around 2011. At the time, drug testing was mainly done either on cells in culture (usually in the form of 3D spheres or organoids) or on animals—which both missed the complexities of a real human tumor. The device I had in mind would help doctors decide which treatment is most effective because a biopsy is the most faithful representation we have of the tumor outside of the patient. Biopsies are only available in scarce quantities, making microfluidic technology—with its tiny micro-channels—a good match for this challenge. Fortunately, we found a third collaborator also at the University of Washington—an enthusiastic neurosurgeon, Bob Rostomily, who had already begun to work with slices of brain tumors (glioma) in culture.

I realized that, like in Huh's device, the key to keeping the intact slices alive was a porous membrane. One of the most successful methods to keep slices alive is to grow them on top of a porous membrane with air above and fluid below, an approach introduced almost three decades ago and since applied to many types of tissue slices, ranging from brain to different cancers.[22] I reasoned that we could grow our cancer slices on top of the membrane and deliver microstripes of drugs that would diffuse through the membrane from a network of microchannels below. Huh's device only had two inlets and two outlets, but we wanted to test dozens of different drugs. My student Tim Chang found inspiration in ninety-six-well plates and bonded the microchannels underneath the multi-well plate so that each well was used as an inlet to our device (figure 6.9).[23] The first tests are usually run with mice tissue, either healthy brain slices (figure 6.10) or human cell–derived tumors inoculated into mice (called xenografts), because they are more homogeneous than patient tumors. The engineering merit of this design—which we nicknamed OncoSlice—is that it is intuitive to use: much like the electronics in the dashboard of a modern car, the microfluidic channels are hidden from the user. My wife Lisa Horowitz became its lead developer in the Rostomily lab. (So yes, we have a lot of OncoSlice discussions at home, especially when she is cooking and I'm cleaning the dishes next to her.) Our lab now fabricates the device entirely in plastic by laser cutting[24] or CNC milling.

In April 2018, Heidi Kenerson, a research scientist in our collaborator and cancer liver surgeon Ray Yeung's lab, handed off to Lisa some liver slices of metastatic colon cancer for OncoSlice experiments. In his research, Dr. Yeung specializes in the study of liver cancers. Many of these patients end up dying. We are always both sorry and immensely grateful for these patient volunteers who give us part of their tumor. They know it is too late to help them but still volunteer the tissue altruistically. That

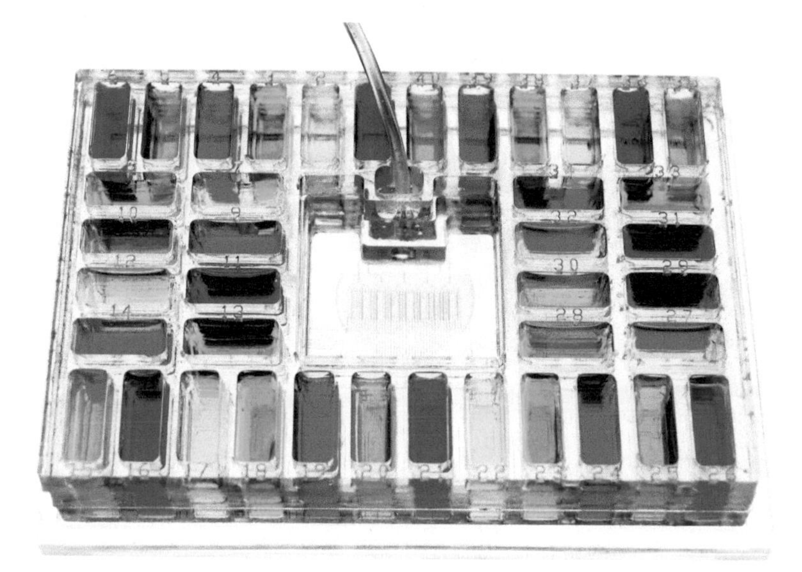

Figure 6.9

Microfluidic OncoSlice platform for multiplexed delivery of drugs to intact tissue slices. The device is made in PMMA by laser cutting. The sample is placed in the center portion of the device. In a drug testing experiment, the dyes are substituted by noncolored solutions containing drugs. The total area of tissue exposed to drugs at the center of the device is about 2 centimeters by 1 centimeter.

Image courtesy of Adán Rodriguez and Albert Folch, University of Washington.

day, the large size of the tumor allowed Dr. Yeung to extract multiple centimeter-sized tumor chunks for Heidi to slice and put in culture. We are lucky the operating room is a five-minute walk from the Yeung lab, and the Yeung lab is another three-minute walk from our lab. Lisa placed the petri dishes on ice in a Styrofoam container and brought them to our lab. There she carefully placed thc slices in the center of the OncoSlice device for a two-day treatment with drugs, including two different standard drug cocktails, and was able to measure the efficacy of several treatments simultaneously. We have also repeated a similar procedure for brain cancers (with Dr. Rostomily) and pancreatic cancers (with Dr. Venu

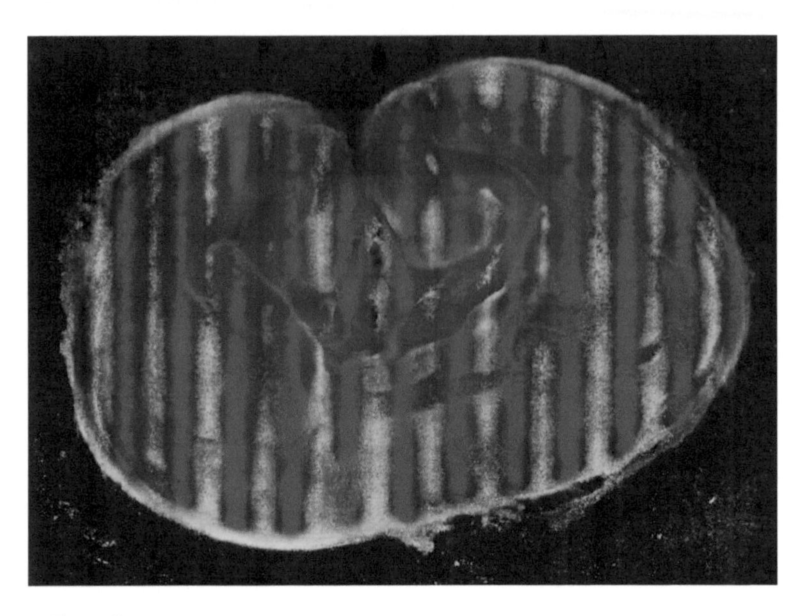

Figure 6.10
A brain slice after exposure to live dyes in a stripe pattern with the OncoSlice platform. Each strip is approximately 100 microns wide.
Image courtesy of Tim Chang and Albert Folch, University of Washington.

Pillarisetty).[25] The study was published in *Nature Precision Oncology* in 2020.

Other labs are using the spheroid approach or taking different cell types and mixing them with tumor cells. Many tumors cannot be sliced easily (they are rather mushy) and, with proper microfluidic design, the experimenter can choose how to grow the cells in a configuration that is optimal for perfusion and microscopy. A collaborative effort between the labs of Steve George at the University of California, Davis and of Abe Lee and Chris Hughes at UC Irvine has built vascularized tumor-on-a-chip devices that permit robust growth and large-scale screening of tumor cells (figure 6.5).[26] Introduction of immune cells such as macrophages triggered various physiological responses against (as well as in defense of) the tumor cells, which shows that the model was performing realistically. A team led by Russell Jenkins at Dana-Farber Cancer Institute and Roger Kamm at MIT minced tumors into very small fragments and placed them inside hydrogel-containing microfluidic devices to study the reaction of native immune cells to the application of immunotherapy drugs.[27] The lab of Thomas Gervais at Polytechnique Montréal and ours have developed techniques to immobilize arrays of live tumor pieces that are microdissected into regular sizes so they can fit in microfluidic traps for selective drug testing.[28] All these microfluidic approaches retain much of the human tumor microenvironment and could not only lead to cheaper drugs but also, in principle, provide results rapidly enough, within days of surgery, to guide the choice of effective initial therapies.

* * *

Given what is at stake, it is not surprising that a growing number of companies are offering their organs-on-chips to pharma, each with a different technology, patented chip design, or organ. The three oldest companies in this space—Hurel Corporation, Hepregen, and Organovo, founded in 2007—are all from the United States. InSphero (Switzerland, 2009), TissUse (Germany, 2010), and Mimetas (Netherlands, 2011) were started shortly afterward. Nortis (Seattle, 2012) and Emulate (Boston, 2013, by Don Ingber) followed, and after them half a dozen startup companies are fighting to grow in this exciting cauldron. Thanks to organs-on-chips, the day when most new drugs will be developed and tested directly (and only) using human tissues is fast approaching.

7 THE PREGNANCY TEST AND THE RISE OF PAPER MICROFLUIDICS

The pregnancy test is unique in more than one way. Its result can bear good or bad news, depending on the situation. "Yes" can mean happiness or trouble. "No" can mean the end of anxiety or the end of hope. I have had the same test result in my hands on two different occasions, and the same result meant very different things to me at those times in my life. When I was eighteen, the "No" was a relief to my girlfriend and me. I did not even know I was holding a microfluidic device; I had just started my physics studies, and I had never heard of microfluidics. Fifteen years later, on a different continent, when I had already started my microfluidics lab at the University of Washington and my wife was trying to get pregnant, multiple repetitions of the same "No" became a source of frustration instead. Finally, at last, the second band lit up: "Yes!" Nine months later, our first child was born.

Another unique characteristic of the test is that it is based on clever microfluidic principles. The power of the test to speak directly to its users derives from the microfluidic engineering hidden inside. It is so user-friendly and portable that anybody can operate it and read it, anywhere, without batteries—capillarity magic at your fingertips. The pregnancy test detects a human hormone with an antibody printed onto a strip of paper inside the test. But if the antibody is changed—say, for an antibody against the COVID-19 virus—it becomes a COVID-19 test. It's that simple and beautiful.

THE PREGNANCY TEST

The home pregnancy test has been dubbed "a private little revolution."[1] A woman's desire to know whether she has become pregnant after intercourse is ancient and has many motivations: whether the pregnancy is voluntary or not, a woman knows that it—or its absence—will have consequences that will be very visible and lasting. The information provided by the pregnancy test is empowering: it allows one to make decisions about one's body, one's future, friends, and family. Sometimes it is a matter of life or death. This information helps plan whether to continue the pregnancy, to test for abnormalities in the fetus, or simply to abstain from drinking a glass of wine—so early knowledge is vital. The home pregnancy test is now a widespread commodity that can be purchased for $10 at any pharmacy. Less well known is that it is a microfluidic device made with a paper strip and that its dissemination ended the suffering of thousands of frogs and rabbits.

Women have been seeking pregnancy tests for more than three thousand years, but have only gained access to them recently. Egyptian doctors would ask women to urinate on barley and wheat: "let the woman water [them] with her urine every day with dates [and] the sand, in two bags. If they [both] grow, she will bear. If the barley grows, it means a male child. If the wheat grows, it means a female child. If both do not grow, she will not bear at all."[2] "Piss prophets" in the Middle Ages claimed to diagnose pregnancy by visually examining the turbidity of a woman's urine. There is indeed something in the urine of pregnant women: high levels of a hormone called *human chorionic gonadotropin*. The name is hard to remember so everyone refers to it by its abbreviation, hCG. But hCG does not affect the urine's turbidity and there is not enough to make it a reliable plant-based assay.[3]

The first reliable pregnancy test was devised in 1927 in Berlin by German scientists Selmar Aschheim and Bernhard Zondek. In the Aschheim–Zondek test, or A-Z test, researchers injected human urine into immature female mice for a few days. Then they sacrificed the mice to measure whether their ovaries had grown to a larger size than average from exposure to the woman's hCG; "yes" meant that the woman was pregnant, but the test took a few days. In 1931, Maurice Harold Friedman and Maxwell Edward Lapham at the University of Pennsylvania improved the test with rabbits.

Rabbits were easier to inoculate (typically in their marginal ear vein). The results were available in twenty-four hours; women could get their results in a week if the results were sent by mail, but sooner if they were telegraphed.[4] The rabbit test, or Friedman test, gave rise to the expression "the rabbit died" for a positive pregnancy test, although the rabbit was sacrificed regardless. In 1933, two students in Lancelot Hogben's lab at the University of Cape Town in South Africa announced that they had worked out a pregnancy test with frogs instead of rabbits.[5] Hogben was the British entomologist who developed the African clawed frog, known as *Xenopus laevis*, as a model organism for biological research. The frog test—wrongly attributed to Hogben later—was much faster and cheaper than the rabbit test because the labs did not need to sacrifice the animals and could thus reuse them: if they injected pregnancy urine (typically mailed into the lab) under the skin of a female frog, the animal—induced by hCG—started laying eggs after twelve hours, a process that did not harm the frog and that was visible to the naked eye. This animal-based testing became widely used for about three decades, until the 1960s, when assays based on antibodies became available.[6]

Not long after Yalow and Berson published their immunoassay technique in 1959 (see box 5.3),[7] anti-hCG antibodies became available. It became possible to detect pregnancy by performing a simple hCG immunoassay in a test tube with a woman's urine. In 1967, Organon Pharmaceuticals, a company in New Jersey, hired Margaret Crane as a freelance graphic designer to work on a new line of cosmetics. In the lab, she saw some test tubes and was told that they were for a pregnancy test, the so-called hCG immunoassay. The test involved mixing two components to produce a color-changing chemical reaction that lasted two hours and detected hCG in the urine sample.

"Could a woman use that at home?" Crane wondered. She designed a set of tubes that would hold the necessary solutions and be easy to use at home. "I was absolutely certain this product would be very useful. That a woman should have the right to be the first to know if she was pregnant and not have to wait weeks for an answer," Crane said.[8] Organon was hesitant at first, but finally decided to patent and market Crane's product. The product, which cost $10, was named the Predictor and was sold first in Canada in 1971; it was not sold in the United States until 1977 due to sexual morality concerns and (male) doctors arguing that women would not be able to cope with the results without them.[9]

A few years later, a microfluidic device made of paper—the famous test with two lines—enabled women to run their hCG immunoassays at home.

PAPER AS A MATERIAL

Paper is inexpensive and readily available everywhere. It is flexible and biodegradable and can be disposed of by burning. Researchers can chemically modify it, and its white color provides a suitable background for colorimetric reactions. Most importantly, its porous structure generates wicking, which acts as a capillarity pump[10]—no additional equipment is required to drive the flow. Paper's wicking property has been as transformative for microfluidics as the invention of the mobile phone was for telephony. Scientists need to tether microfluidic devices made in silicon, glass, PDMS, and plastics to a pump—gravity is included here—in order to power the flow into the channel; they also need a bulky microscope to visualize the results: more than a lab-on-a-chip, it's a chip-in-a-lab. Some devices can function by clever capillary filling designs,[11] but they require many optimizations. Paper can also be used as a sample-preparation sieve to filter and separate particles. In a paper microfluidic device, the channel *is* the pump *and* a filter, so the device becomes fully mobile.

As opposed to mobile telephony, this new mobility comes with a hardware simplification, which facilitates high-volume manufacturing and cost reduction. If you think that electronic glucometers are cheap—they sell on Amazon for $25–$50 for the reader and fifty plastic strips—consider this: they cost the equivalent of a month's food supply in some rural communities in Africa where $1 can buy a few dozen tomatoes or a dozen eggs. There are places where cost is everything.

THE FIRST PAPER STRIP TESTS

Microfluidics is often associated with the microfabrication techniques developed in the 1970s. The word *microfluidics* does not appear in the scientific literature until 1992,[12] but researchers were tinkering with microfluidic phenomena for centuries well before the field got its name. One of the earliest developments in analytical chemistry consisted of detecting chemicals with paper strips impregnated with reagents. Both the detection and the reagent impregnation utilized the microfluidic wonder now referred to as *capillary wicking*—the spontaneous drawing of fluid into the cellulose fibers by capillary action. Pliny the Elder—the Roman naturalist who died in Pompeii during the eruption of Mount Vesuvius in 79 CE—wrote that "to detect adulteration with shoemaker's black [ferrous sulfate], place a portion on papyrus previously steeped in extract of gallnuts which blackens immediately in the presence of the adulterant."[13] The seventeenth-century Anglo-Irish physicist Robert Boyle, who is widely regarded as the first modern chemist, observed that violet syrup—which women at the time commonly concocted as an ailment for various pains[14]—changed to a green color when a basic solution was added and to a red color if an acidic solution was added. From this observation, Boyle devised the first chemical indicator: a strip of paper soaked with violet syrup that would turn green or red depending on whether it was dipped in a base or an acid, respectively—an invention that Boyle and many others after him extended to various plant juices.[15] In the late nineteenth century, Edme-Jules Maumené in France and George Oliver in Britain developed urine tests based on the impregnation of a porous substrate such as wool or paper to detect glucose.[16] Their tests used a chemical reaction that was not very specific to glucose because other sugars and acids present in urine could also reduce the reagents. In the early 1930s, G. Gutzeit[17] and Fritz Feigl[18] realized that the capillary properties of filter paper were useful for enhancing colorimetric spot reactions. In these reactions, they delivered a drop of sample solution to paper impregnated with the appropriate reagent. Herman Yagoda improved this assay by the addition of wax printing in 1937.[19] Ralph Muller and Doris Clegg used Yagoda's wax patterning technique in 1949 to make fluidic circuits on paper[20] for separating chemical compounds (that is, *paper chromatography*[21]).

The first consumer products for urine testing became commercially available in the late 1950s. In 1957, Alfred and Helen Free at the Ames Corporation developed the Clinistix, a urine reagent strip impregnated with the enzyme glucose oxidase, peroxidase, and ortholidine;[22] in the presence of glucose, the ortholidine was specifically oxidized into a deep-blue product. The British pathologist Joachim Kohn noticed that one could also use the Clinistix to detect glucose from a drop of blood.[23] Based on this finding, Ernie Adams at Ames developed the Dextrostix, a platform that included an optical reader and strips to measure the amount of sugar in the blood. At the same time, Boehringer Mannheim developed the Chemstrip bG, also to detect glucose in the blood. Different versions of these urine and blood paper strips were used—some with optical readers—until the late 1990s for self-monitoring of diabetes, growing to a multibillion-dollar point-of-care market.[24]

THE CLEARBLUE

One should expect that companies such as Siemens, IBM, and Hewlett-Packard, who were developing computing equipment, would race to develop the inkjet printer. Yet the story of why and how Unilever, a large manufacturer of soap and margarine, decided to develop the first microfluidic pregnancy test based on a paper strip is a bit more intriguing.[25] Founded in 1929, Unilever had become by the 1950s a vast conglomerate of companies with products ranging from detergents to fish fingers seeking new markets for expansion and diversification. A medical division was started. The division recruited immunologist Philip Porter as part of an effort to strengthen its science base in the 1960s. Porter had a PhD in immunochemistry and must have been aware that researchers had used paper for a while as a low-cost platform for the separation of compounds in analytical and clinical chemistry techniques using strip assays. Porter established the main elements of a stellar immunodiagnostic program at Unilever—including strategic patents—and, together with Paul Davis, developed the groundbreaking dipstick concept of a simple, efficient one-step enzyme-linked immunosorbent assay (ELISA).

The subsidiary Unipath subsequently employed the basic principles of this assay to launch the paper-based pregnancy test known as ClearBlue in 1985. (The universal two-strip pregnancy test is the ClearBlue One Step introduced in 1988, featuring the chromatographic detection of hCG hormone and control strip.) After the i-STAT in 1984, the ClearBlue was the second microfluidic health-care product to market—before researchers had ever used the word *microfluidics* in the scientific literature. Unipath later developed a range of similarly formatted medical diagnostic kits. Despite its great success, Unipath sales amounted to less than 1 percent of Unilever's total,[26] so Unilever sold Unipath to the Inverness Group in 2001.

One of the most powerful features of pregnancy tests is that they can be easily repurposed to detect pathogens such as viruses. Inside the pregnancy test (figure 7.1)—the first of a class of microfluidic tests now called *lateral flow assays* (box 7.1)—you will find a simple strip of paper that hides a marvel of immunoengineering. The strip contains three sensing lines, each impregnated across the paper with different antibodies. The three lines are called the reaction line, test line, and control line, of which only

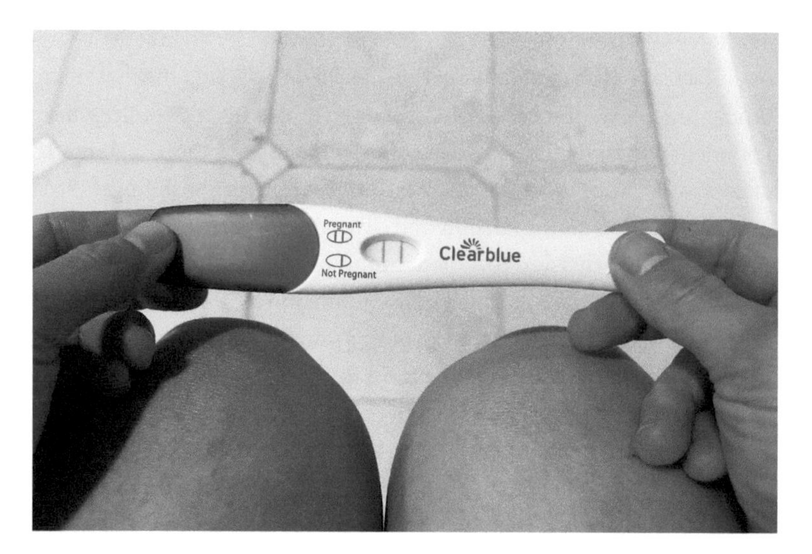

Figure 7.1
Pregnancy test. The paper microfluidic strip is visible through the window. For this test to work, the blue cap must first be lifted and the left end must be dipped in urine. The control line on the right of the strip is visible, indicating that the test worked properly. In this case, the test line is also visible, which indicates that the woman is pregnant. Note that the reaction line (to the left of the test line) is hidden inside the plastic device.

two (test and control) are visible to the user. The antibodies on the test and control lines are chemically linked to the paper, whereas the antibodies on the reaction line are free to flow. The antibodies against hCG in the reaction and test lines are different: they each have been engineered to recognize a different part of the hormone. The test line antibody, on the other hand, is designed to recognize the reaction line antibody itself (not the hormone). There are two other important components that make the test visible to your eyes. The reaction line antibody is chemically linked to an enzyme. This enzyme is a protein that reacts with a dye that has been impregnated into the test and control lines. Under certain conditions, the enzyme is brought into contact with the dye, triggering a color reaction that is visible as a band.

BOX 7.1
Lateral Flow Assays for Everything—Including COVID-19

Lateral flow assays are now used ubiquitously in clinics and labs to test urine,[27] saliva,[28] sweat,[29] serum,[30] plasma,[31] whole blood,[32] and environmental water, among other fluids. By changing the antibodies, researchers can detect specific antigens,[33] antibodies,[34] and products of gene amplification.[35] They can also screen for animal diseases,[36] pathogens,[37] toxins,[38] and water pollutants,[39] among others—including COVID-19 (figure 7.2). The assays can differ in the sample introduction method, in the color-generating reactions, in the number of lines, and in the method of activation of the control line (some assays incorporate a bit of antigen).

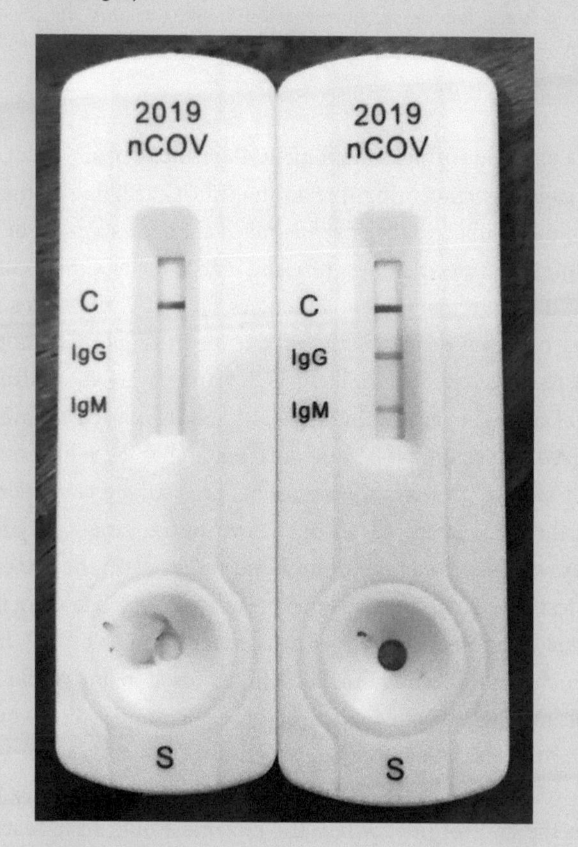

Figure 7.2
A lateral flow assay for COVID-19. The test detects two antibodies generated by exposure to the virus. The test on the left is negative and the test on the right is positive.
Image courtesy of Dr. Janusz Rat.

(continued)

BOX 7.1 (continued)

During the recent COVID-19 pandemic, several companies have raced to develop lateral flow assays to detect coronavirus in a drop of blood, either by detecting the antibodies generated by the virus in the body or by directly detecting a viral protein. Quidel has released its Sofia SARS Antigen Fluorescent Immunoassay, which uses an immunofluorescence-based lateral flow assay to detect the presence (or not) of nucleocapsid protein from the SARS-CoV-2 virus. Cellex, ChemBio, and several other companies[40] were able to quickly deploy COVID-19 lateral flow assays for the detection of SARS-CoV-2 antibodies in blood by changing the antibodies of existing assays.

When a user adds urine containing hCG hormone upstream of the reaction line, capillary action brings the hormone hCG with fluid to the reaction line. The fluid solubilizes the enzyme-linked antibodies against hCG in the reaction line; the antibodies recognize and bind to the hCG molecules in the urine, but there are many more antibodies than hCG molecules so many antibodies continue downstream without an hCG partner. As the fluid arrives at the test line, the antibody–hCG partners are recognized by the other set of antibodies against hCG. Thus, the hCG molecules have become immobilized in a sandwich between two antibodies. Now the enzyme can react with the dye present in the test line, producing a coloration. Meanwhile, as the urine keeps advancing, it carries the excess enzyme-linked antibodies, which arrive at the control line and perform the same reaction as in the test line. That's the two-lines-means-yes result shown in figure 7.1.

If, instead, the user had added urine containing no hCG hormone upstream, the antibodies against hCG in the reaction line would not have been able to partner with hCG. Therefore, the antibodies would have flowed past the test line without any binding (or reacting of the enzyme), but they would have ended up all the same being captured by the antibodies in the control line, where the enzyme would have reacted with dye and triggered a coloring reaction. That would have been a one-line-means-no result.

DIAGNOSTICS FOR ALL USING MICROPATTERNED PAPER

George Whitesides had been championing PDMS for a few years but was worried that it could not be applied to run diagnostic tests in rural settings and remote villages of developing nations. As a chemist, he appreciated the operational simplicity of the pregnancy test. However, the lateral flow assay is binary—it is designed to provide a qualitative (yes/no) answer. For many diagnostics, doctors want to run a quantitative assay that answers how sick the patient is. The development of such complex assays—involving mixtures and timing of reagents—demands the compartmentalization of fluids in separate volumes, either by creating chambers in glass or PDMS, or by patterning paper (figure 7.3). In contrast, the pregnancy test and paper electrophoresis devices are based on a single rectangular strip—a single fluidic compartment of limited functionality.

Running cheap diagnostic assays by patterning paper had been on George Whitesides's mind for half a decade, but the project would not materialize until PhD student Andrés Martínez joined his group. Martínez was born in Oakland, California, because his Bolivian dad had been sent to California in the 1970s to be shielded from the brutal political repression in Bolivia. Martínez's parents met in California and joined the hippie movement in Berkeley, but decided to go back to Bolivia when Andrés was five so their four children could get a better life and education in the countryside. "I grew up raising chickens and harvesting fruits and vegetables," remembers Martínez. At eighteen years of age, Martínez left home for a community college east of San Francisco, from which he transferred to Stanford to graduate in chemistry. Then he was admitted to Harvard for a PhD in 2004, planning on doing research on organic chemistry. "When I arrived at Harvard, to be honest, I didn't know who Whitesides was," Martínez told me in his charming Bolivian Spanish with a big laugh, opening his eyes in disbelief at the confession he had just made. But the Whitesides lab had a nice offering of pizza and beer during open house, and Whitesides presented a project he called "Simple

Figure 7.3
Paper-based biosensors for the detection of amitriptyline, an antidepressant drug. The channels are made of nitrocellulose using femtosecond-laser ablation. *Image courtesy of Andreas Dietzel, Technische Universität Braunschweig.*

Solutions" that struck a chord in his farm-raised Bolivian soul: the idea of developing diagnostic systems that (as eloquently put by Whitesides in a TV interview of the time) "can work in economically constrained environments—systems that have to be affordable, workable in a low-resource setting, and must produce actionable information."[41] Martínez instantly thought: "This lab is for me." George Whitesides offered him a PhD slot on the spot and set him up to work with biologist Samuel Sia. Sia was doing postdoctoral research on PDMS diagnostic systems for low-resource settings[42] and was working closely with Vincent Linder, another postdoc (box 7.2).

Linder left the group in April 2004 before Martínez arrived in the fall of 2004, and Sia finished his postdoc in the winter of 2005, six months after

BOX 7.2: BIOGRAPHY
Vincent Linder

Linder was born in 1975 in Neuchâtel (Switzerland), a small university town in the heart of the watch industry. The area, with the small-scale machining expertise devoted to watchmaking, had become a fertile ground for microtechnology. The University of Neuchâtel built a world-class MEMS and microfluidics program led by Nico de Rooij. Linder studied chemistry at Neuchâtel. After graduating in 1998, the microtechnology department of the University of Neuchâtel offered him a multidisciplinary PhD project that focused on surface chemistry, biochemistry, "and a new thing called microfluidics" under the advisory of de Rooij. When he finished his PhD, he wanted to expand to the interface with biology, so he reached out to Whitesides, who accepted him for a postdoc starting in April 2002.

Sia had started in the lab in 2002 at the same time as Vincent Linder. "I had the great fortune to meet Vincent Linder, who taught me all this microfluidics stuff," says Sia. In the Whitesides lab, Linder and Sia worked together developing PDMS-based diagnostic devices for improving global health,[43] which served as the foundational work for Claros Diagnostics, the diagnostics and global health company they cofounded.

Martínez joined. At this point, it would become clear to Martínez that Whitesides had other plans for him. Shortly after Sia left, Martínez met with his PhD advisor once again. "PDMS microfluidics is very complicated, it needs a lot of infrastructure," Whitesides pointed out in that meeting. "We need something simpler. Why don't you try paper channels," he added.

Whitesides suggested Martínez buy a few types of hydrophobic solutions so he could pattern the paper by impregnation. For a few months, Martínez tried solutions that make surfaces and clothing impermeable, such as Scotchgard, but the techniques he used for patterning were very rudimentary or manual. "I was using pens, and I ruined two or three printers," he recalls with a happy tone in his voice. Martínez tried using vacuum to create suction and also vapor deposition, to no avail. He was growing frustrated, so toward the fall of 2005, he met again with Whitesides and showed him his results. "It's a good start, but we need high definition to get visually appealing, reproducible patterns," said Whitesides. The two sat down and drafted a list of classical high-resolution patterning techniques that Martínez should try next, such as screen printing and photolithography. Martínez recalls putting away the list. In early winter 2006, Martínez found the list of patterning techniques under a bunch of books on his desk. He decided to give SU-8 photolithography a try.

There was one problem. Martínez was thinking of using chromatography paper, a type of paper made of pure cellulose that has been used by chemists for a long time to filter and separate solutions. But the clean room facility at Harvard did not allow this paper because it sheds particles—paper is not "clean." Martínez then made a decision that would have made his dad proud: to break Harvard's rules. He would enter the facility at 5 a.m. when nobody else was working to bring in paper without being seen. Martínez had the instincts of a true pioneer: one cannot start a revolution and be a rule-follower at the same time. Once inside the clean room, he poured SU-8 on his chromatography paper pads and allowed the photoresist to impregnate the paper. Then he developed it as one would develop SU-8 on a silicon wafer, and he observed the SU-8 pattern magically appear under the yellow light of the clean room—only the SU-8 pattern of channels was embedded inside the paper, not above the surface. Back in the lab, as he tried to fill the SU-8-walled paper channels with a colored dye, he noticed that the paper would not wick—somehow, the parts of the paper that he had exposed to SU-8 had become hydrophobic. Martínez then thought of making it hydrophilic by exposing the paper to an oxygen plasma, and *then* it worked! Later this problem was solved by developing with solvents like acetone (different from the recommended SU-8 developer, which must have been leaving some residue). He used the technique to demonstrate 3D channels in paper (figures 7.4 and 7.5) for the simultaneous detection of glucose and protein in 5 microliters of urine and published it in 2007 in *Angewandte Chemie*,[44] a paper that has accumulated more than 2,600 citations by 2021. Of course, the paper itself was a public acknowledgment of his rule-breaking. When I asked him if, in all these years, anybody at Harvard had questioned how he had managed to pattern paper in a clean room, Andrés Martínez—now a respectable professor, but still looking like the only young student that would undertake the paper microfluidics project—laughs again: "Nobody noticed!"

Unbeknown to Martínez, Vincent Linder had already patterned paper with SU-8 in March 2004, a few weeks before leaving the group. Linder planned and did those experiments with Sam Sia. "We thought we should use the paper that is routinely used in clean rooms and that has large pore sizes," explains Linder referring to the TechniCloth lint-free paper that

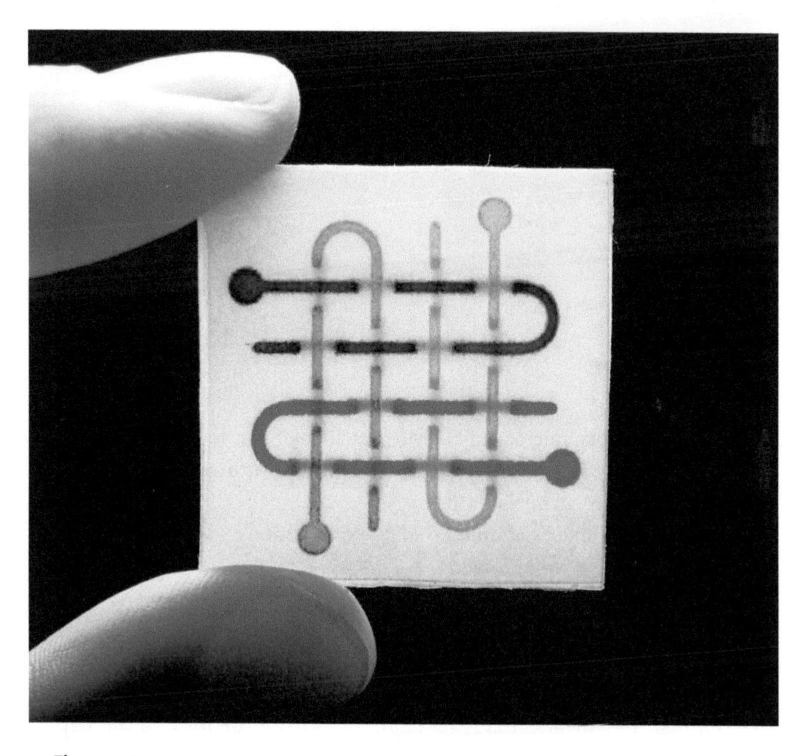

Figure 7.4
Multilayered paper microfluidic device showing a braided 3D network of four nonintersecting microchannels.
Image courtesy of Andrés Martínez, California Polytechnic State University.

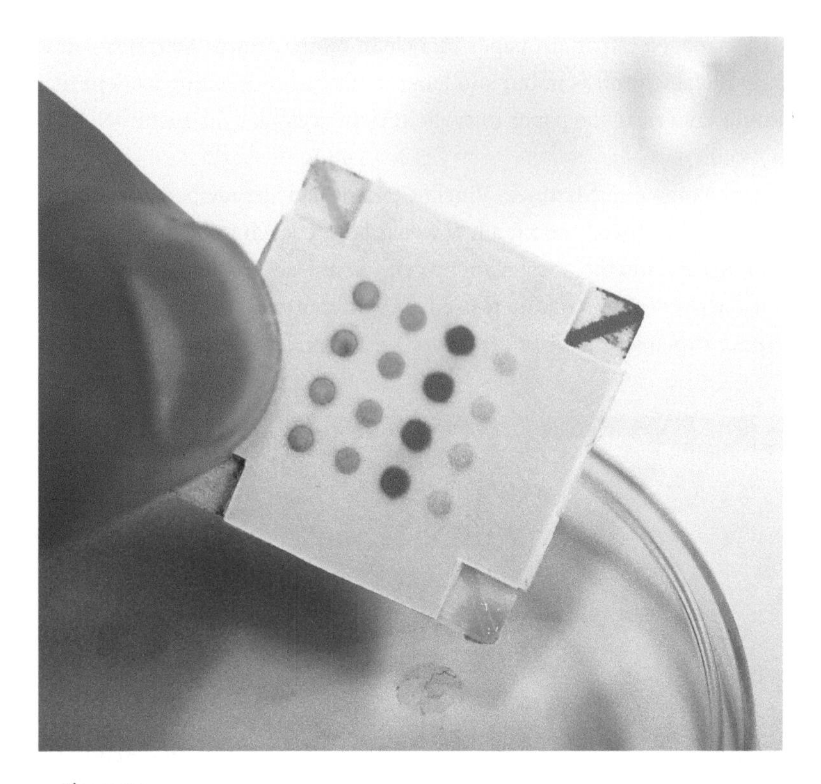

Figure 7.5

Multilayered paper microfluidic device. Each color dye is loaded in a different corner of the square and connects to four dots through a 3D network of paper-based microchannels.

Image courtesy of Andrés Martínez, California Polytechnic State University.

is 45 percent cellulose and 55 percent polyester. (The chromatography paper used by Martínez, on the other hand, is 100 percent cellulose and is not lint-free.) “We were interested in patterning the bulk of the material, not just the surface, to create channels within the bulk of the paper, by closing some pores selectively through polymerization of SU-8 in the pores according to the layout of a photomask,” Linder adds. “The initial idea that George had was to pattern SU-8 on paper the same way you pattern SU-8 on silicon, but we could not get it to work. Linder tried, but the

SU-8 would not stick. Then Linder soaked the paper in the photoresist, and that worked," remembers Sia, who also recalls reporting the results to Whitesides at some point.

To this day, Martínez does not know whether Whitesides had forgotten about Linder's results or whether he was trying to stimulate his creativity by making him think in the right direction when he gave him the list of things to try in the winter of 2005. We do know that Whitesides had been trying to convince people in his group for many years and that he was not able to convince Sia to follow up. "We did stuff on paper because George kept asking, but Vincent and I felt that monolithic systems had more horsepower," says Sia. (By "monolithic systems," Sia means "standard microfluidics.") In January 2006, Martínez called Byron Gates, an ex-postdoc of the lab, to ask him a question about self-assembled monolayer chemistry. During the conversation, Martínez mentioned that he was working on paper. "Ah, George has finally been able to convince someone to work on paper! He has been trying for years!" exclaimed Gates. Gates is not the only one to have those memories. Daniel Chiu explains that halfway through his postdoc, around 1999, Whitesides took him to the cafeteria across the street from the lab to grab some "lunch"—"usually coffee and half a muffin for George"—as he would sometimes do with individual members, and "tried to get me to think about paper analytical devices." Given the size of the group and how busy Whitesides was, this type of individual discussion was not frequent for a lab member, but "George always had interesting perspectives and lots of ideas, so I tend to remember those conversations well and take them to heart," emphasizes Chiu. "George mentioned that thin-layer chromatography [a form of chemical separation technique that is similar to paper chromatography] is used all the time in chemistry and is very convenient and provides good information, and the principle was really just wicking in a substrate and easy readout," recalls Chiu. "He said I really ought to think about it," says Chiu. But Daniel Chiu had other plans and, like many others afterward, decided to work on PDMS instead.

Whitesides must have grown frustrated by his vision being rejected from within his ranks, and waited for years with monk-like patience until the right traveler stopped at his Harvard monastery. I do not know any other scientist with the tenacity to postpone a project for five or six years, knowing very well that their idea was simple enough to be conceived by someone else. Martínez, finally, fit all the requisites—and Whitesides was ready to collect the fruits of his resolute persistence.

After the *Angewandte Chemie* paper in 2007, the Whitesides group rushed to expand the capabilities of the technique. Martínez applied the method to make 3D paper constructs, this time with lint-free paper allowed in clean rooms—published in *PNAS* in 2008[45]—and presented a cell-phone-based assay in *Analytical Chemistry* that same year.[46] Paper ELISA immunoassays,[47] as well as programmable devices based on paper push buttons[48] and origami folding to incorporate temporal control,[49] were soon developed by Martínez and other members of the group. In 2009, Bingcheng Lin's group and postdoc Emanuel Carrilho in the Whitesides group—within a few months—were able to bypass the clean room altogether by developing a wax printing approach for printing paper microfluidics.[50] If someone noticed the violation of clean room protocol in the first manuscript, there is no record of it—and Harvard will not make a fuss now about a few, long-gone lint particles given the wealth generated by this research in later years. Soon large numbers of researchers started using these techniques, precisely because they were so inexpensive and simple to implement.

The impact of Andrés Martínez's work cannot be overstated. As of 2020, Martínez's six first-author publications from his PhD have been cited more than eight thousand times, with an insanely high total production of fourteen PhD publications. By the time Martínez finished his PhD in 2010, complex mobile biochemical assays in resource-poor settings were no longer a dream. The professor who was once a farm boy had made it possible for microfluidic assays to reach remote farming communities worldwide.

* * *

Many microfluidic engineers must have wished they had come up with Martínez's paper micropatterning technique ten years earlier. We had

had it in our hands when we were looking at pregnancy tests as we were expecting children, but we must have been thinking about something else. Whitesides's scientific vision and leadership cannot be ignored here. Whitesides was the person who first articulated the advantages of both PDMS and paper for microfluidics. Having proposed PDMS, he was also the first to propose the alternative of paper. He first understood that the advantageous properties of paper (low cost, wicking) were the solution to some of the disadvantages of PDMS (tethered system, poor manufacturability) and articulated the message for the microfluidics community. Whitesides handed researchers the Holy Grail of microfluidics: a microfabrication technology that can be used seamlessly for prototyping *and* manufacturing—so the work done in their laboratories could be immediately disseminated and commercialized.

The simplicity of paper made it easy to shift from PDMS to paper. Not surprisingly, troves of researchers—mostly from the field of sensors—started developing point-of-care mobile sensors based on paper microfluidics, often using techniques other than photolithography or printing, such as laser cutting (figure 7.6). In 2007, Whitesides started Diagnostics For All, a company founded "with a shared commitment to saving lives and alleviating disease in developing countries and other resource-poor settings through low-cost, innovative, practical diagnostic devices."[51] In 2009, Chuck Henry's group utilized carbon and silver/silver chloride inks to write a three-electrode design in paper and demonstrate the electrochemical detection of glucose, lactate, and uric acid in human serum.[52] Paul Yager's group implemented a clever paper-based valve based on dissolvable paper that, when dissolved by the addition of the sample, released a spring-loaded mechanical lever (figure 7.7). The lever release was used to deliver the reagent that starts a very sensitive colorimetric reaction for the detection of influenza virus.[53] The large (and growing) number of researchers now using paper microfluidics is a tribute to the low cost, accessibility, and translational power of this technique: everything that was done in the lab could be immediately applied because the manufacturing constraints were so low.

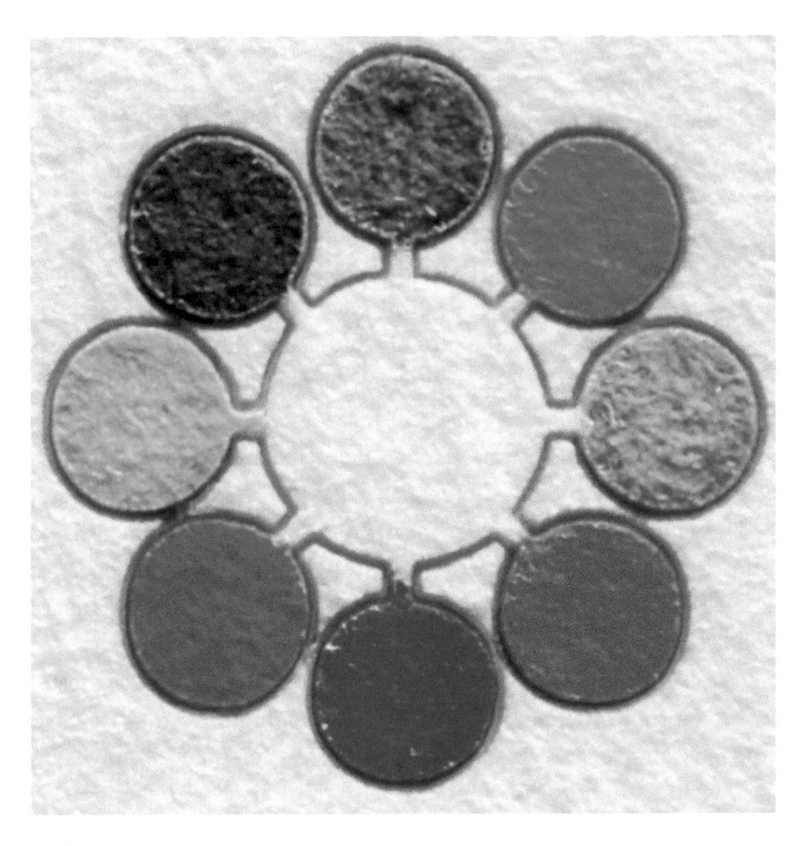

Figure 7.6
Microfluidic diagnostic device with eight reaction zones laser-cut in chromatography paper preloaded with various reagents. The sample (e.g., a urine drop) is added in the middle. The total width of the device is ~7 millimeters. *Image courtesy of Md. Almostasim Mahmud and Brendan D. MacDonald.*

Some will argue that everything comes at a price. Compared to PDMS- or plastic-based platforms, assay sensitivity and specificity in paper microfluidics are not improved,[54] and automation schemes[55] and cell-based studies[56] (while possible) are both impaired relative to PDMS. "In paper microfluidics, you start with a low-performance, low-cost system, and then you try to improve the performance, but most of the time, you will complexify it and take away the low-cost advantage of paper," says Sia. "I'd

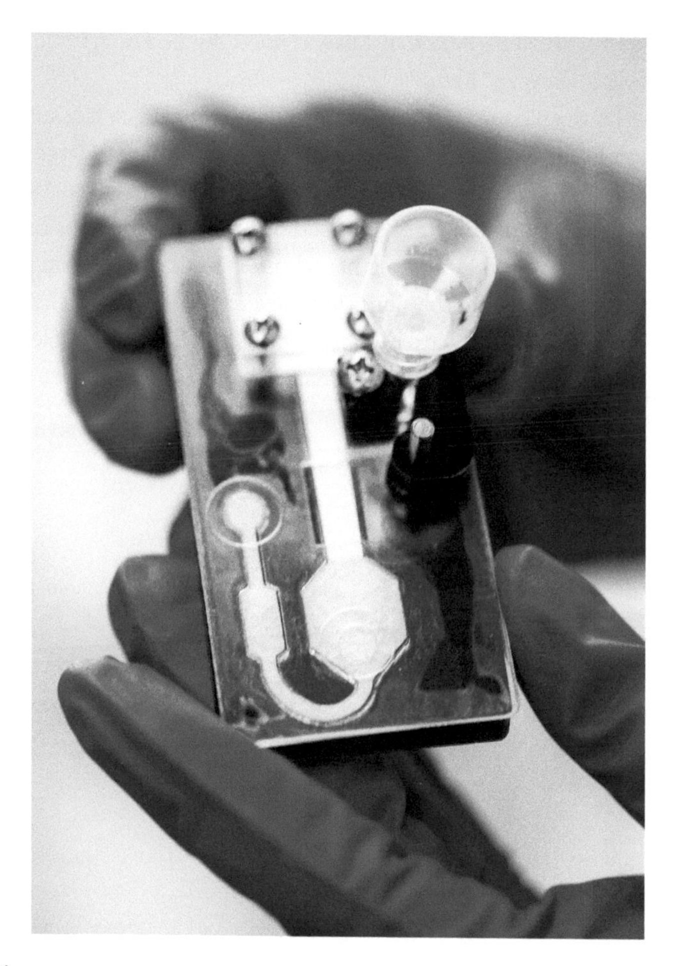

Figure 7.7
Microfluidic diagnostic device for detection of influenza virus. A reagent for the detection of the virus is introduced into the plastic cup. When the sample (introduced on the circle on the left) advances by capillarity to the square (between screws), it dissolves a piece of paper, releasing a spring-loaded lever next to the base of the cup. The lever releases the pressure on the base of the cup, allowing for the reagent to flow and mix.
Image courtesy of the University of Washington.

rather start with something that's like a 'microfluidic Ferrari' and work on making it cheaper," he adds.

In the bigger picture, the global health field will have to face an even more significant challenge. Diagnostics For All initially raised large sums of funds from foundations and federal sources, but ultimately dissolved. The cruel truth is that resource-poor settings are often too poor to even afford $1 devices—and that fundamental problem demands geopolitical rather than technological solutions. But the efforts of these paper microfluidics pioneers will not have been in vain—after all, many people compromise on a Toyota for many daily tasks. When top performance is not required, cheaper is always better.

8 IN SEARCH OF EXTREMELY RARE CELLS

Sometimes, there are a few cells in our blood that should not be there (figure 8.1). Their extremely low densities—only a few rare cells among more than a billion blood cells per milliliter—makes their detection exceedingly challenging. The ability to detect these rare blood cells, such as cancer cells in the blood of patients in the early stages of metastasis, could improve our health.

Around 2000, Harvard bioengineering professor Mehmet Toner (box 8.1) wondered why such cells had not been detected efficiently enough in cancers other than very advanced metastatic cancer. On a large piece of paper, he summarized all the rare-cell detection techniques used until then. This list included technologies such as antibody-coated magnetic beads and shooting cells down through capillaries (flow cytometers). He realized that people were tweaking techniques not intended for rare cells. "It's like racing against a Formula 1 with a Hyundai—you are not gonna win," he says with one of his usual proverbial comparisons, pausing to make sure I have assimilated the full meaning of his words.

Turkish-born and speaking with a polished Turkish accent, Toner is one of the few scientists who is a world leader in two disparate disciplines—cryobiology (the study of how living organisms survive low temperatures)

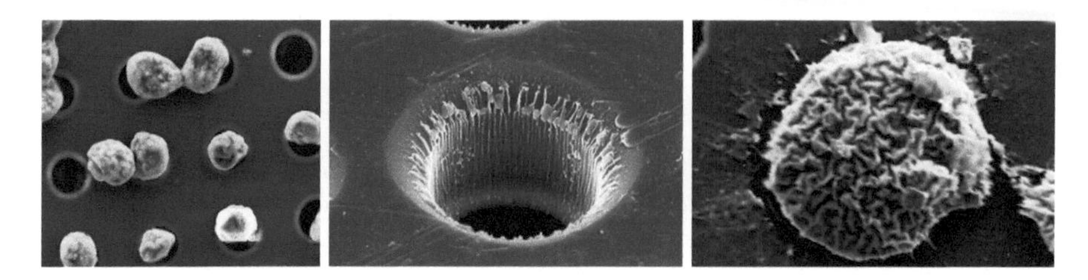

Figure 8.1
Scanning electron micrographs of circulating tumor cells trapped using a size-selective sieve (a membrane containing holes) made of parylene (left). The holes are 10 microns in diameter (center). All the other blood cell types have passed through the holes, but the circulating tumor cells are about 15 microns in diameter and do not fit through the holes (right).
Images courtesy of Yu-Chong Tai, California Institute of Technology.

and microfluidics. His air of a blue-eyed Buddhist monk lends his comments the hint of wisdom from a far-off land. "I also realized that they had come up with the hypothesis that these [rarer] cells do not exist. My alternative hypothesis was that the technology sucked," concludes Toner with a mischievous smile.

The explosion of the field of *circulating tumor cells* (*CTCs*) after Toner's group developed a set of microfluidic chips for isolating CTCs with high efficiency testifies that his hypothesis was correct. Toner envisions that his ultra-high-throughput microfluidic chips will soon enable the collection of CTCs from large blood volumes in the clinic. He believes that "we all have cancer during our lives" and that our bodies' immune system can fight cancer away most of the time. However, these CTCs represent early signs of cancer cells that have evaded the immune system. He hopes that the early detection of these culprits—invisible to traditional techniques but not his microfluidic systems—will become the new frontier of cancer therapy.

CIRCULATING TUMOR CELLS

One of the cruelties of cancer is that exposure to it is statistically inescapable: most people will suffer the loss of a loved one to cancer if they

BOX 8.1: BIOGRAPHY
Mehmet Toner

Mehmet Toner was born in Istanbul in 1958. His father was a renowned Turkish architect who built one of the largest hospitals in Ankara. By his own account, Toner was "not a good student" in high school and wanted to be a surgeon initially but could not get into medical school. He started taking his studies seriously when he entered Istanbul Technical University to study mechanical engineering in 1977. Toner fondly recalls professors Aksel Ozturk and Abdurahman Kilic as "wonderful people who motivated me and got me hooked on solving problems and understanding things." By the time he graduated in 1982, he had published two papers on heat transfer and was at the top of his class. Accepted for a master's in mechanical engineering at MIT, Toner landed in Cambridge, Massachusetts, in 1983. He realized there was a new field that was starting at the time called biomedical engineering. "I started reading about it with great interest because of my failure in medicine," says Toner. He joined the lab of Ernest Cravalho, who was working on the thermodynamics of freezing cells—a heat transfer problem applied to biology. One year later, Toner moved to the Health Sciences & Technology (HST) Program (an MIT–Harvard joint PhD program in medical engineering). There he took medical courses and clinical rotations in addition to his regular PhD work. Around 1989, Toner helped Ron Tompkins and Maish Yarmush raise funds to start the Laboratory for Surgical Science and Engineering, later renamed the Center for Engineering in Medicine. After finishing his PhD in 1989, and a short postdoc also at MIT, he obtained his present Harvard faculty position at Massachusetts General Hospital (MGH) in 1990. In a few years, Toner built a research program that is incredibly successful at both cryobiology—a field in which he is still very active—and microfluidics.

do not experience the disease themselves. Doctors will diagnose an estimated 1.8 million new cancer cases in the United States in 2020, and 600,000 patients will die from it[1]—making cancer the second leading cause of death after heart disease.

Cancer patients rarely die of the primary cancer. More often than not, the primary cancer spreads to a distant, vital organ in a process called *metastasis.* It is this secondary cancer that ends up causing death. Researchers have investigated the cellular mechanism for this deadly spreading for a long time. In 1869, Thomas Ashworth, a resident physician at Melbourne

Hospital, Australia, looked at the blood of a patient with metastatic cancer under the microscope and observed non-blood cells that were morphologically similar to those in the primary tumor. In his study, he concluded that these CTCs must have traveled from the original tumor site to thirty secondary distant sites.

More than one and a half centuries later, we know that tumors shed cells into the blood and lymphatic vessels. Once in circulation, they randomly extravasate at another body site, forming *micrometastases*—most of which are taken care of by our immune system in their regular housekeeping functions. However, if one or more of these escapes the immune system, the tumor can spread. The seeding process can occur at various stages of tumor development, so cells at secondary sites may range from dormant to aggressively malignant.

We cannot stop metastasis yet, but doctors can take a blood sample and look for CTCs—a test called a *liquid biopsy*.[2] The detection and analysis of CTCs can help doctors find cancer at an early stage, plan treatment, and find out how well treatment is working or if cancer has come back. Doctors can repeat this procedure over time to understand what molecular changes are taking place in the tumor. The number of CTCs before (and sometimes during) treatment is a good predictor of the chances of survival of patients with metastatic breast,[3] colorectal,[4] and prostate[5] cancer. In patients with breast cancer, *clusters of CTCs* (even more rare than single CTCs) appear to have twenty-three- to fifty-fold increased metastatic potential.[6]

But finding CTCs can be as tricky as finding a needle in a haystack. CTCs are extremely rare. In a patient with metastatic disease, every milliliter of blood has approximately *one CTC* surrounded by a few million white blood cells and *a billion* red blood cells. So far, there is only one FDA-approved method for CTC detection—brand-named CellSearch (box 8.2). One problem finding CTCs with CellSearch has been that it relies on the CTCs expressing a molecular tag on their surface called *epithelial cell adhesion molecule* (*EpCAM*). Some cancers, like triple-negative breast cancer or melanoma, have very low EpCAM expression—making them undetectable by CellSearch. Additionally, the CellSearch procedure

BOX 8.2
CellSearch

CellSearch is an FDA-approved method for CTC capture based on magnetic-activated cell sorting (MACS). CellSearch separates CTCs from as little as 7.5 milliliters of blood. The method uses 100-nanometer-sized, superparamagnetic nanoparticles conjugated to an antibody against the epithelial cell adhesion molecule (EpCAM). Thus, the nanoparticles effectively tag the CTCs that express the EpCAM surface marker. As the magnetic particle–decorated CTCs go through a column surrounded by permanent magnets, the magnetic field captures the CTCs. CellSearch provides a CTC count doctors use to evaluate the prognosis of patients that have been diagnosed with breast, colorectal, and prostate cancer.[7] Researchers use CellSearch to capture various types of cells—as long as an antibody against a surface marker for that cell is available.

entails processing steps such as centrifugation, washing, incubation, and nanoparticle decoration that can potentially damage the CTCs.[8] In the end, it captures very small cell numbers. Furthermore, an ideal CTC assay involves not just counting the CTCs but also keeping them intact and alive for further analysis, including drug responsiveness.[9] Here is where microfluidics can help, explains Mehmet Toner.

THE CIRCULATING TUMOR CELL CHIP

In early 2000, as I was getting ready to leave the Toner lab and applying for faculty positions, Mehmet Toner started thinking about the then-extraordinary challenge of capturing rare cells. At the time, state-of-the-art microfluidics focused on basic devices for biological and biochemical analysis that seem simple by today's standards. Yet Toner was interested in clinical applications and the analysis of "bodily dirty fluids," as he calls them (e.g., blood, plasma, urine, or central nervous system and peritoneal fluids). Mehmet remembers being inspired by the rare-cell literature of the time, when researchers were struggling to capture rare cells because they flowed them through a capillary-based sorter or poured large quantities of fluid into a magnetic-bead sorter (MACS) like CellSearch. "I wondered about something in between," he thinks back on his inspiration. He reasoned that, whenever you need to look at rare events, you need to look at large volumes. But microfluidic devices would clog. To process billions of particles with single-cell precision, he needed the equivalent of "the computer mainframe for microfluidics."

But the applications in cancer were not easy to define because there were still uncertainties in the technology, biology, and clinical applicability. Toner realized he would need help. In December 2003, Toner and Ron Tompkins, then the director of the Center for Engineering in Medicine, grabbed their coats and crossed the street. They carefully trod the snow on the ground, walked under the glass awning of Navy Yard Building 149—a military-coded building that had been used initially as a shipyard storehouse and was now the site for MGH's Cancer Center—and entered the building. They took the elevator to the seventh floor and knocked at the door of the late Kurt Julius Isselbacher, the legendary gastrointestinal doctor who had been at MGH since the 1950s; Isselbacher was expecting them. It is impossible not to admire Isselbacher, a compassionate doctor who had fled Nazi Germany in 1936 as an eleven-year-old Jewish boy. He spent the last fifty years of his life building bridges and connecting dots at MGH, giving advice behind his half-lowered glasses with a benevolent, thick-lipped smile and wise, puffy eyes. As the director of the MGH Cancer Center since its creation in 1987, in his last year before stepping down, Isselbacher knew everyone and was the most responsible for its excellence. Toner describes him as "a Jedi" for his role as "a mentor for everyone at Mass General." "Your first application could be the EGFR mutation that Haber has found," said Isselbacher, referring to his MGH oncologist colleague Daniel Haber.

Luckily for Toner, Isselbacher already knew in late 2003 that the Haber lab had identified a critical mutation in lung cancer. The papers eventually came out in the *New England Journal of Medicine* and *Science* in the late spring of 2004.[10] In the lung and many tissues, the mini-protein epidermal growth factor (EGF) is essential for cell growth and division.

EGF activates the cells via a molecule called the EGF receptor (EGFR) hanging out of the cells' membranes. Lung cancer is divided into two broad types: small-cell lung cancer—the most aggressive type—and non-small-cell lung cancer—the most common type of lung cancer, which is slower growing and accounts for almost 90 percent of cases. Most often, it causes few or no symptoms until the cancer is in an advanced stage. Haber and his team had found that in non-small-cell lung cancer the EGFR gene had suffered a mutation and was too active. In other words, these tumor cells tended to grow because they were too sensitive to EGF. Haber's group was able to inactivate the receptor specifically by blocking it with a small-molecule drug called gefitinib. This drug became the first application of targeted therapies in lung cancer. There were targeted therapies for other cancers known in 2004, but researchers had already explored them at length so the potential for new discoveries was low. Isselbacher was merely suggesting to Toner that partnering with Haber could be mutually beneficial in the research path to explore the EGFR mutation.

Indeed beneficial it was. In May 2004, Toner and Tompkins went back to the Navy Yard Building 149, but this time to see Haber, the new director of the MGH Cancer Center. Daniel Haber had not worked on CTCs, but he immediately found Toner's ideas very appealing. As a doctor, Haber saw several potential advantages in using the devices envisioned by Toner. He knew that an X-ray could miss tumor nodules below a specific size (on the order of a millimeter), so Toner's device might be able to catch the disease at earlier stages. A method based on detecting CTCs in the blood would bypass the risk, discomfort, and cost of an invasive needle biopsy. A lung needle biopsy requires hospitalization, an operating room, and anesthesia—none of which are required to draw blood for a CTC count. Perhaps most importantly, Haber was also aware that tumors often become resistant to therapy. A method based on blood sampling would allow for serial monitoring of the disease while the patient was undergoing therapy. Instead of taking a single diagnostic picture of the tumor at the beginning of the treatment, they would be able to obtain a "molecular movie" of the cancer as it evolved during treatment. It could be revolutionary.

They quickly sketched a work plan. Microfabricating the chip was not straightforward. Toner tried to convince Sunita Nagrath, who had joined the lab with a PhD in computational fluid mechanics, to work on a different project. However, she insisted on working on the CTC chip, which, to her credit, required fluid dynamics modeling to optimize CTC capture. The final design consisted of an equilateral triangular array of microposts, with each post 100 microns tall and 100 microns in diameter and a gap of 50 microns between microposts. Marty Schmidt suggested the use of deep reactive ion etching to plasma etch the microposts in silicon. The post array was sealed with a transparent tape layer to form a chamber. For the detection of CTCs, they decided to use an antibody against EpCAM because 90 percent of all solid tumors originate in the epithelial lining (except melanoma and few others), so they all have EpCAM—and blood cells do not have it. In 2007, Nagrath et al. presented in *Nature* a microfluidic device featuring 78,000 EpCAM antibody-coated silicon posts that could capture, isolate, and analyze CTCs from a blood sample.[11] The device isolated an average of 132 CTCs per milliliter of blood from all tested patients with metastatic cancers (non-small-cell lung cancer and prostate, pancreas, breast, and colorectal cancers), but not from healthy controls. Having accumulated more than 3,800 citations by 2021, the 2007 Nagrath paper is now being cited at the highest rate in the history of microfluidics. This paper and those that followed have generated a frenzy of research on CTCs, with many labs presenting new microfluidic designs in various conferences every year.

In 2008, a follow-up paper in the *New England Journal of Medicine* by the Toner and Haber team used the microfluidic silicon-post device to detect mutations in the EGFR gene in circulating lung-cancer cells, opening the possibility of monitoring changes in epithelial tumors during treatment.[12] The vast majority of the EpCAM-positive cells turned out to be, indeed, CTCs. However—to make Toner's task more difficult—they found that some cells shredded their EpCAM marker, and others got covered with platelets, which caused the CTCs to miss the antibody-coated posts. So they decided to build a chip that worked based on negative selection—it eliminated all cells that were EpCAM-negative—and then analyzed the few remainders. "We pull out all the cells that are not supposed to be in the blood, and then sequence their DNA to find which ones are cancerous," explains Toner. Sometimes CTC cells come in clusters, so a modification of the posts into a triangular-post array allowed for capturing such clusters.[13] In prostate cancer patients, these CTC clusters correlate with high patient mortality. "As a therapy, we want to break down the clusters by targeting the junctional protein that holds the clusters together," says Toner. Compared to CellSearch's sensitivity, which only detects CTCs in 20 percent of patients with advanced metastatic disease, Toner's CTC chips reach detection in 80–100 percent of patients depending on the cancers. The diagnostic test only requires detection of a few (three to five) CTCs from 20 milliliters of blood (a human has an average of 5 liters of blood).

Haber and Toner realized early on that they were facing a very complicated problem. "This can't work as a collaboration, we need to join forces," they both reasoned. So they got ready as for an extended camping trip like two good old pals. Their research is now so intertwined that they no longer collaborate the old way—they have *joined* their labs. Here is an enduring legacy of Isselbacher's. Students can work in either lab without barriers. Toner and Haber write their grants and do all their budgeting jointly. "He is very funny, we have so much fun together," Toner says of his thick-mustached, cheerful colleague and friend Haber. The success of the CTC device led the National Institutes of Health to award Toner with a BioMEMS resource center that employs thirty-five people and, as of 2020, can boast of having produced 152 microfluidics publications and twenty faculty jobs since its inception in 2012.

Other applications opened up and the competition to commercialize them became fierce. Toner shrugs away the possibility that some of these companies might infringe on his intellectual property, answering with one of his Confucian proverbs: "We poked the sleeping bear." There are now about ten companies that work on microfluidic rare-cell searching, but that has also created an ecosystem that is richly innovative and beneficial to all. As an example, Toner's lab applied a microfluidic chip to the problem of prenatal diagnosis to detect fetal cells in the mother's blood, in part because it seemed more tractable than cancer: in prenatal diagnosis, both the clinical indication and application are clear—if you can find the cells, you can look, for example, for chromosomal abnormalities or Down syndrome, and solve the medical problem. When a fetus has suspected abnormalities, the doctor typically recommends an amniocentesis. In this procedure, the doctor uses a needle to sample the amniotic fluid surrounding the fetus through the mother's belly to obtain fetal cells and amniotic stem cells. However, an amniocentesis has a miscarriage risk of about 1 percent. Since amniotic fetal cells are found in the mother's blood, and in larger quantities than CTCs, Toner thought that a microfluidic test would make for a safe replacement of amniocentesis. For this work, they decided to use a silicon-post design that, instead of capturing the cells like Velcro as in the CTC chip, deflected the cells laterally based on size.[14]

The principle of this device, originally reported in 2004 by James Sturm and Bob Austin at Princeton, is called *deterministic lateral displacement* (*DLD*)[15] and may well be one of the cleverest uses of laminar flow. Maternal red blood cells (RBCs) and platelets are approximately 2 microns in size, whereas nucleated red blood cells (NRBCs, from the fetus) and white blood cells (WBCs) are larger (around 5–10 microns). To understand how the device works, imagine a set of regular streamlines as wide as a cell flowing around the posts. The post array acts like a size filter where the pore size of the array is dictated not by the spacing between posts but by the width of the flow stream, which is smaller than the post spacing. When whole blood was passed through the post array, cells that were small with respect to the flow stream (RBCs and platelets) drifted along the flow and stayed in the stream. In contrast, cells that were large compared to the flow stream (NRBCs and WBCs) were displaced out of the stream and deflected sideways when they encountered a micropost. Final separation of the NRBCs from the WBCs was achieved by application of a magnetic field. With the help of Diana Bianchi—a world expert in prenatal diagnosis—the research was spun rapidly into a company. Sold to Illumina in 2013 and renamed as Verinata Health, the company presently sells its tests around the world.

The Toner group did not stop at silicon-post designs. The silicon-post approach is deceptively simple compared to the swirly designs developed later that incorporated channels with ornate twists and wondrous curves, each performing a specific function. The mastermind architect of those shapes was Dino Di Carlo, a young microfluidic engineer who joined the Toner lab in 2006 (box 8.3). Di Carlo would revolutionize microfluidics with his ingenious designs that take advantage of cells flowing at dizzying speeds, which was very useful to Toner's "mainframe" vision of processing billions of cells in an hour. Because the inertia of the fluid starts playing a role at those speeds and causes many eerie phenomena to appear, we specifically refer to that high-fluid-speed mode of operation as *inertial microfluidics*.[16] If the microfluidics field were a set of races in which devices compete, inertial microfluidics—a discipline on its own that has many applications beyond detecting CTCs—would be the equivalent of the Formula 1 races.

INERTIAL MICROFLUIDICS

Di Carlo humbly attributes to "serendipity" his first observation of inertial microfluidics. There were many CTC experiments going on in the Toner lab, with people trying to improve the adhesion of CTCs to surfaces so they could trap cells with higher efficiency. Di Carlo started thinking of centrifugal effects that could bring the cells closer to surfaces where they could interact. "If I flow cells fast enough around a curved structure, cells are denser, they are going to get pushed against the wall," he remembers reasoning, mostly about density differences, at the time. Di Carlo acknowledges he did not know about Dean flow back then.

In the fall of 2006, he tested a design he had put into a postdoc fellowship application. "I started testing the design with labeled cells, and the first thing I notice is that the cells are not going to the walls at all, they

BOX 8.3: BIOGRAPHY
Dino Di Carlo

Dino Di Carlo has the name of a movie star, and he could pass as one—you will never miss him among a group of scientists. He is the very blond, tall guy with the neatly unshaven look and shaded glasses who moves smoothly among the crowd. His parents met in the Detroit Italian community. His father was a language teacher that emigrated to Michigan in search of opportunities from Abruzzo, the mountainous south-central region of Italy that lies between Rome and the Adriatic Sea. Italian writer Primo Levi once said that the beauty of Abruzzo and the character of its people are best described as "forte e gentile"[17] (strong and kind)—adjectives that suit Di Carlo perfectly. His mother was born in Michigan because her parents had emigrated, also from Abruzzo, to Detroit to work in the automobile factory. Born in Ohio in 1981, Dino Di Carlo's family moved from Ohio to Monterey, California, for his father's job when he was only six months old. His father had grown up in rural Italy and was fond of gardening, so Dino grew up surrounded by a little farm of pigeons, rabbits, and chickens. His mother was learning to be a nurse when he was in middle school. "The human anatomy books that she was bringing home are the reason I became interested in medicine," Di Carlo recalls. Then his father took a position as lecturer at UC Berkeley, so Di Carlo went to high school and did the rest of his schooling in Berkeley.

After high school, he was admitted into the prestigious UC Berkeley bioengineering program. As an undergrad, he joined Luke Lee's group in 1999 and stayed in his lab for a PhD starting in 2002. Di Carlo describes his days in Luke Lee's lab with intensity: "There are two things I learned from Luke: One is how you present science through images—through iterations and iterations; the other is that he is incredibly creative—he would question the orthodoxy of science. Why is something believed to be true?"

As a student leader of one of the student-run clubs on campus, he got to know one of the women he had met in the undergraduate bioengineering program, also a student leader in the same club, and they got married. In 2005, two months after his first child was born ("my wife was not happy"), Luke Lee sent him to a Gordon Conference at Oxford. There he met Daniel Irimia, who was already working with Mehmet Toner at the Center for Engineering in Medicine at MGH, and invited him to apply for a postdoc. In summer 2006, Di Carlo turned down a postdoc offer from Sandia National Labs in favor of MGH because of the clinical connection.

start focusing to the center of the channel (figure 8.2)—very bizarre!" he says. But he adds: "I realized that it was quite important," not realizing that this is the part that takes anything but serendipity. Di Carlo started testing fundamental effects, like ramping the Reynolds number by changing the flow rate, or comparing it to a straight channel. "Then I started seeing that even in a straight channel I saw three streams, with the middle one twice as bright," he explains. At first, he did not understand all the details, but he was resolved to do so. The middle stream twice as bright turned out to be two streams, one on top of the other when observed on a confocal microscope.

Di Carlo went back to the fluid mechanics literature to understand what was going on. How could cells be focusing without additional forces? What are the intrinsic fluid dynamical forces that act on particles in flow? He found that in 1961 G. Segré and A. Silberberg had reported in *Nature*

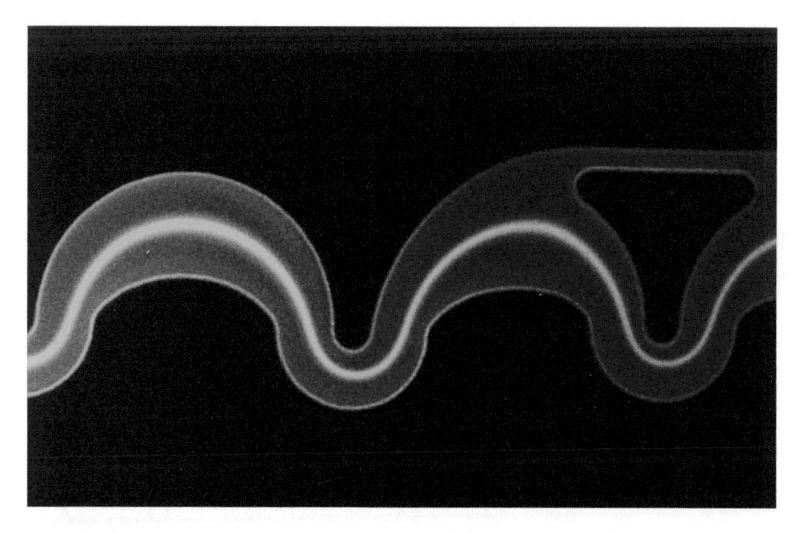

Figure 8.2

Spontaneous focusing of cells (here visible collectively as a fluorescent green trace) to the center of a meandering microchannel, a foundational finding of inertial microfluidics.

Image courtesy of Avanish Mishra and Mehmet Toner, Harvard Medical School.

that larger (millimeter-scale) particles in wider (centimeter-scale) pipes were displaced and spontaneously ordered into an annulus at high flow rates.[18] "So I thought this must be the same effect," remembers Di Carlo. He also learned about the so-called *Dean flow* in microchannels,[19] which appears when channels have a curvature and transversal recirculation vortices occur to conserve the momentum.[20] Dean flows play essential roles in arterial blood flow and various instruments, such as dialysis devices and heat exchangers. Crucially, Di Carlo used high-speed digital cameras to take still frames of the particles. And, against all preconceived notions that the particles would be flowing in random arrangements, he saw that even in the absence of any curvature, the particles were flowing as a regular lattice of pearls enjoying a ride on a microscopic subway (figure 8.3). "The first time I saw those ordered structures, I thought it was amazing," he recalls.

The explanation for the particle ordering effect, which Di Carlo contributed to formulate, is now in textbooks. The parabolic profile of the flow in the microchannel (always faster at the center) drives the particles toward the channel walls. However, the flow around the particles also produces a pressure build-up that prevents the particles from getting close to the walls, thus reaching equilibrium particle distributions.[21] It's as if all the bullets from a machine gun magically organized themselves in the air and hit a target in just four locations, forming a perfect square design. The "magic" that causes the microfluidic ordering is a set of fluidic interactions between the microchannel walls, the particles, and the fluid itself.

Di Carlo reported in 2007 in *PNAS* a microfluidic device that ordered and sorted particles using fast flow.[22] He knew that the next step was to apply these findings to biological particles such as cells. The following

Figure 8.3
High-speed micrograph of the spontaneous ordering of particles inside a microchannel due to cell-wall interactions in inertial microfluidics.
Image courtesy of Dino Di Carlo, UCLA.

year, Di Carlo demonstrated the filtration and separation of platelets from diluted blood at throughputs on the order of a milliliter per minute,[23] a speed comparable to that of macroscale filtration. In 2009, Di Carlo went back to his native California to start a faculty position at UCLA. Some of the key design motifs in these papers still pop up in Toner's group's devices more than a decade after he left.

Toner and Haber immediately noticed that Di Carlo's inertial microfluidics designs could separate particles without labeling them first with magnetic nanoparticles—one of the drawbacks of the CellSearch technology. Would it be possible to separate cells in two stages? The first stage would be to "debulk" the blood of the millions of red blood cells from the WBCs and CTCs by DLD and inertial microfluidics (which is size-selective); the second stage would be to apply antibody-labeled magnetic beads and a magnetic field to perform a separation step that distinguishes the EpCAM-positive from the EpCAM-negative CTCs. In the Toner lab, Emre Ozkumur designed an automated platform—dubbed CTC-iChip (figure 8.4)—that achieves CTC separation in three steps.[24] In the first step, WBCs and CTCs (prelabeled with magnetic beads) are separated from all the other blood components (e.g., red blood cells, platelets) by a DLD post array. In the second step, the output of the DLD array is connected to a Di Carlo–style wiggle channel that focuses the WBCs and CTCs into a single-cell line. In the last step, a magnetic field is applied to separate the CTC cells (which are magnetically labeled) from the WBCs. The CTC-iChip achieved the separation of both EpCAM-positive and EpCAM-negative CTCs in clinical samples and was published in 2013 in *Science Translational Medicine*.[25]

Di Carlo's work inspired many other approaches. Notably, Jongyoon Han and his PhD student Majid Warkiani, leading a Singapore–MIT team, in 2014 created a spiral microfluidic design that causes CTCs to separate by size due to two added effects.[26] First, the lift forces observed by Di Carlo bring the cells to the usual equilibrium positions. In addition, for the same reason that the outer wheels of a car spin faster than the inner ones when it takes a turn, due to the curvature of the channel, vortices appear in the channel to maintain the momentum. Larger and smaller

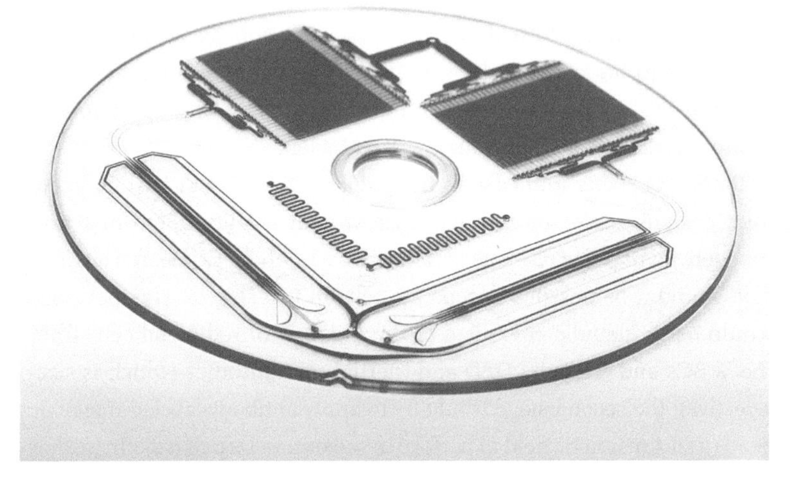

Figure 8.4
The monolithic CTC-iChip (here, filled with pink dye for visualization purchases) is engraved in a 12-centimeter-diameter plastic disk that contains millions of micron-scale features such as the posts forming a DLD array. It processes more than 15 million blood cells per second to obtain highly purified CTCs.
Image courtesy of Mehmet Toner, Harvard Medical School.

cells separate to different ends of the vortex that forms across the channel. To increase the separation effect of these so-called Dean's vortices, the authors fabricated a slanted channel (taller on the outside than on the inside of the curve), further forcing the smaller cells to the outside and keeping the CTCs on the inside. They successfully isolated and recovered more than 80 percent of the cancer cell line cells spiked in 7.5 mL within eight minutes and detected 3–125 CTCs per milliliter in all ten patient samples tested.

* * *

From the sunny office of his Los Angeles home, Di Carlo reveals that "getting the experiments to work was not the main obstacle." When, as a young faculty member, he first submitted a grant on inertial microfluidics, the reviewers did not understand how there could be inertial effects in a microfluidic device and denied Di Carlo's explanations, which had already been published. "The orthodoxy at the time was that there were no inertial effects in microfluidic devices," he remembers somewhat bittersweetly, delivering an Abruzzian message for past and future generations—strong and kind. So Di Carlo and his students put together one mind-boggling demonstration of inertial microfluidics after another. In 2010, PhD student Claire Hur built a simple microfluidic device to count cells at rates on the order of a million cells per second—about three orders of magnitude faster than previous microfluidic cell counters. Albert Mach, Claire Hur, and Elodie Sollier in 2011 developed an oh-so-clever Vortex device that preferentially traps cells of larger sizes while the smaller ones pass by. The vortices are not unlike those mesmerizing water eddies that sometimes form between rocks in a creek,

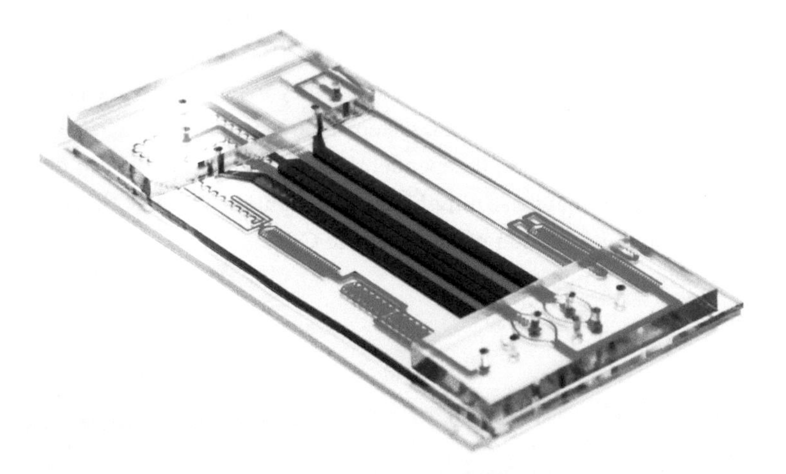

Figure 8.5
Microfluidic platform for ultra-high-throughput, epitope-agnostic isolation of circulating tumor cells from large blood volumes.
Image courtesy of Avanish Mishra and Mehmet Toner, Harvard Medical School.

which trap a leaf and refuse to let it go. After the device has separated the smaller cells, researchers release the larger cells simply by slowing the flow.[27] The Vortex device is now commercially available (Vortex Biosciences, cofounded by Elodie Sollier) and focuses on CTC detection. The technology foundation underlying several other companies (MicroMedicine, CytoVale, Cellular Highways in the UK, and Biolidics in Singapore) utilizes inertial microfluidics.

Di Carlo does not need to worry about convincing anyone anymore—through his work, inertial microfluidics is now the orthodoxy. In a pace of progress that is reminiscent of that of Moore's law for semiconductor chips, the Toner–Haber group continuously pushes the speed and precision of the CTC isolation process. In 2020, Avanish Mishra made further improvements to the CTC-iChip design—also based on inertial microfluidics principles—that made it two orders of magnitude faster (figure 8.5).[28] Importantly, the team sought to avoid labeling the CTCs with antibody-conjugated magnetic beads. This step depends on the presence of the EpCAM epitope (the part of the molecule recognized by an antibody) on the CTCs and could cause the CTC-iChip to miss some CTCs that do not express EpCAM. Hence, they magnetically labeled the white blood cells instead for the last step of the separation—which is now "epitope-agnostic," that is, independent of the presence of EpCAM epitope. Toner hopes that these ultra-high-throughput microfluidic chips will soon enable the collection of highly viable CTCs from ever-larger blood volumes in the clinic.

9 MICROFLUIDICS AND 3D PRINTING

3D printers are to regular printers what sculptors are to painters. They allow materializing designs out of the two dimensions of a plane into the "third dimension." 3D printers started as research tools, but the success and reach of 3D printing have extended beyond research to art, building, haute cuisine, and even children's classrooms.[1] What you might not know is that several types of 3D printers are based on microfluidic principles, and in this final chapter I will tell you about three main types.

You are already familiar with one of them because the book started with it: inkjet printers. The same microfluidic nozzles that shoot droplets of ink on paper can also dispense multiple layers of polymer droplets on a surface. Following the commands of a computer file that contains a 3D digital design, the printhead builds the physical part, droplet by droplet and layer by layer. Different nozzles, fed from different ink reservoirs, are activated in parallel to print multiple polymers at once, or in sequence. Usually these polymer materials are light-sensitive, so as soon as the droplet hits the top layer of the print, the printer turns on a UV light to rapidly cure the droplets in place. However, these inkjet-based 3D printers are still limited in the range of materials that they can print. The polymers cannot recreate materials with the properties of glass or paper, for example, but the printer can generate 3D models with astonishing detail (figure 9.1).

In a chicken-and-the-egg twist, 3D printers equipped with these microfluidic inkjet printheads have been used to print . . . microfluidic devices[2] that exploit both the 3D and the multi-material fabrication capabilities of the technique. For example, Dana Spence and colleagues at Michigan State University printed plastic microfluidic sensors incorporating connector gaskets; the gaskets were 3D-printed in a different, flexible material.[3] In 2016, Liwei Lin and his graduate student Ryan Sochol at UC Berkeley 3D printed microfluidic valves, pumps, fluidic diodes, fluidic capacitors, and fluidic transistors that were assembled into circuits with various automation capabilities and creative 3D shapes (figure 9.2)[4] very unlike the flat-chip architectures of yesteryear. The structures were

Figure 9.1
This miniature reproduction of an industrial complex was 3D printed with a multi-material inkjet printer from a single digital file that specifies all the components.
Image courtesy of HP Inc.

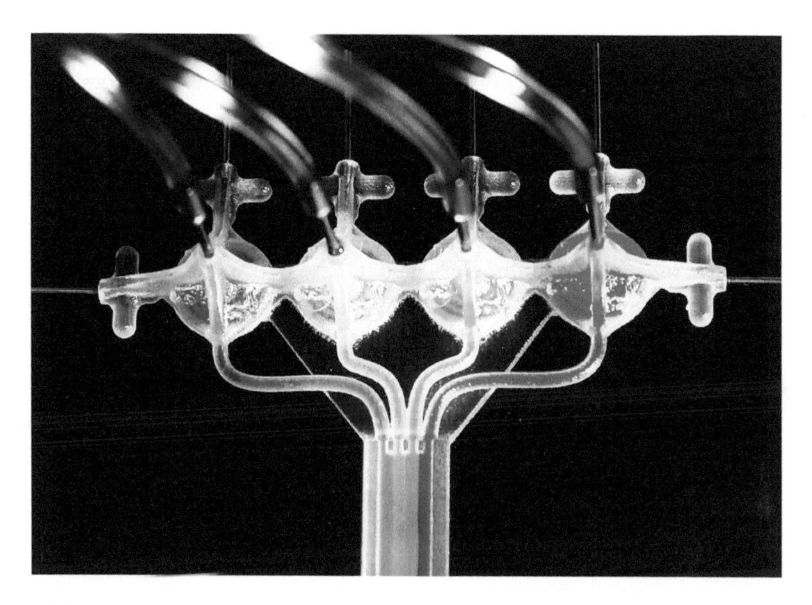

Figure 9.2
Microfluidic device consisting of four valves. The device has been 3D printed with a multi-material inkjet printer from a single digital file that specifies all the components.
Image courtesy of Ashley Tsai and Ryan Sochol, University of Maryland.

printed in two materials, one acting as the walls of the channels and the other (a wax) acting as the filling, which was later dissolved away.

The second type of 3D printer also uses a microfluidic printhead. If you have seen a 3D printer in your children's school, or you own one in your house, it is likely a 3D printer based on the extrusion of a melted plastic filament through a microfluidic (heated) nozzle. The filament extrusion feels much like toothpaste coming out of a tiny tube. This approach, called fused filament fabrication (FFF) and also known by its trademark name of Fused Deposition Modeling (FDM), was invented in 1988 by mechanical engineer Scott Crump. These 3D printers are undoubtedly the cheapest and most user-friendly. The filament is usually fed from a large spool into a printhead that contains the nozzle, which moves in all three *X*, *Y*, and *Z* directions under computer control. A digital design file instructs the printer to build the print by following a programmed path, moving the printhead in *X*, *Y*, and *Z* layer by layer, as it "reads" the file (figure 9.3). The material is chosen by the user—dozens of different plastics and even conductive and fluorescent polymers are available—but the resolution is determined by the nozzle's diameter—between 300 microns and 1 millimeter—and is fixed. Importantly, the widespread availability of FFF 3D printers enables distributed manufacturing approaches. During the COVID-19 pandemic, there was a manufacturing and distribution shortage of personal protection equipment such as face masks for healthcare workers, and thousands of users spontaneously

Figure 9.3
Close-up of the printhead of a fused filament fabrication 3D printer featuring two brass nozzles. The filament is gray-colored and the right nozzle is in the middle of printing an Eiffel Tower.
Image courtesy of Arman Naderi, University of Washington.

shared their designs online and contributed their 3D printers to print these parts for hospitals—all enabled by a microfluidic nozzle.

A third microfluidic 3D printing strategy is called *direct ink writing (DIW)*, a sister technique of fused filament fabrication. DIW is also based on a nozzle that is scanned in *X*, *Y*, and *Z*, but in DIW, the nozzle extrudes slurries or pastes at room temperature. You have probably seen pastry chefs using food 3D printers to build custom cakes with fancy shapes, because chocolate and other foods make good slurries (figure 9.4)—you can even experience it at FoodInk, a restaurant where all the food is 3D printed in front of you. To serve cooked food, Hod Lipson's lab at Columbia University has developed multi-material food 3D printers that are capable of simultaneous infrared cooking.[5] In a similar twist, some artists use 3D printers to print ceramics. In some parts of the world, contractors use cement 3D printers to print an entire neighborhood of houses—each with a different, personalized floor plan—for a small fraction of the cost of what it would have taken to erect a dull array of identical suburban houses. Electrical engineers have miniaturized this concept and are using metallic viscous solutions that are fed from syringes or pressurized reservoirs. More than one reservoir is often operated in parallel and a system of valves is used to switch from one slurry or paste to the next. Researchers have employed DIW to create LEDs,[6] batteries,[7] strain gauges on flexible substrates,[8] antennas,[9] interconnects,[10] and electrodes within biological tissue.[11]

Exploiting DIW's high biocompatibility and multi-material capability, several groups have developed *bioprinters* and *bioinks* composed of hydrogels and cell-laden solutions. PDMS devices can be used as printheads that mix and extrude spatially patterned hydrogels.[12] In 2013, Konrad Walus and coworkers incorporated PDMS valves to add programmable multi-material hydrogel switching capability,[13] an approach that became the basis of the highly successful Canadian company Aspect Biosystems. Mike McAlpine and coworkers custom-built a 3D printer in 2015 that fabricated a multi-material three-chamber microfluidic device, made in polycaprolactone and silicone, to study interactions between neurons, glia, and

Figure 9.4a and Figure 9.4b

Food 3D printing by DIW. (a) 3D-printed and laser-crusted cheesecake about 2–3 inches long. This food product consists of seven printed ingredients: graham cracker paste, peanut butter, jelly, Nutella, banana, cherry drizzle, and frosting. Graham cracker served as a structure component within which to pool the other softer ingredients. The top surface was crusted at the end with a blue laser diode. (b) Printing of a 1–2-inch-wide hexagonal structure using carrot purée. The structure is later filled with thinner liquids.

Images courtesy of Jonathan Blutinger and Hod Lipson, Columbia University.

epithelial cells.[14] Jennifer Lewis's group developed microfluidic printheads for direct ink writing of viscoelastic inks (figure 9.5)[15]—a printer that has now become the envy of the 3D printing world for its precision and multi-material capabilities (figure 9.6). In 2017, Kit Parker and Jennifer Lewis printed cardiac organ-on-a-chip devices with three types of liquid inks—piezoresistive ink, high-conductance ink, and viscous PDMS ink. The device integrated 3D-printed contractility force sensors and enabled the study of drug responses and contractile development of human stem-cell-derived cardiac tissues for more than four weeks.[16]

It was not until 1996, when Jennifer Lewis attended a Gordon Research Conference on Ceramics and met Joseph Cesarano from Sandia National Labs, that she became actively involved in the field of 3D printing (box 9.1). Gordon Conferences require attendees to stay, socialize, and mingle for the whole week in close quarters. Cesarano had coinvented a 3D printing technique for patterning 3D ceramic lattices from a colloidal ink—essentially a particulate paste or slurry. After running into Jennifer at the

Figure 9.5
3D printing with viscoelastic inks. Quickly switching the ink inside the nozzle results in the extrusion of a multi-material thread.
Image courtesy of Jennifer Lewis, Harvard University.

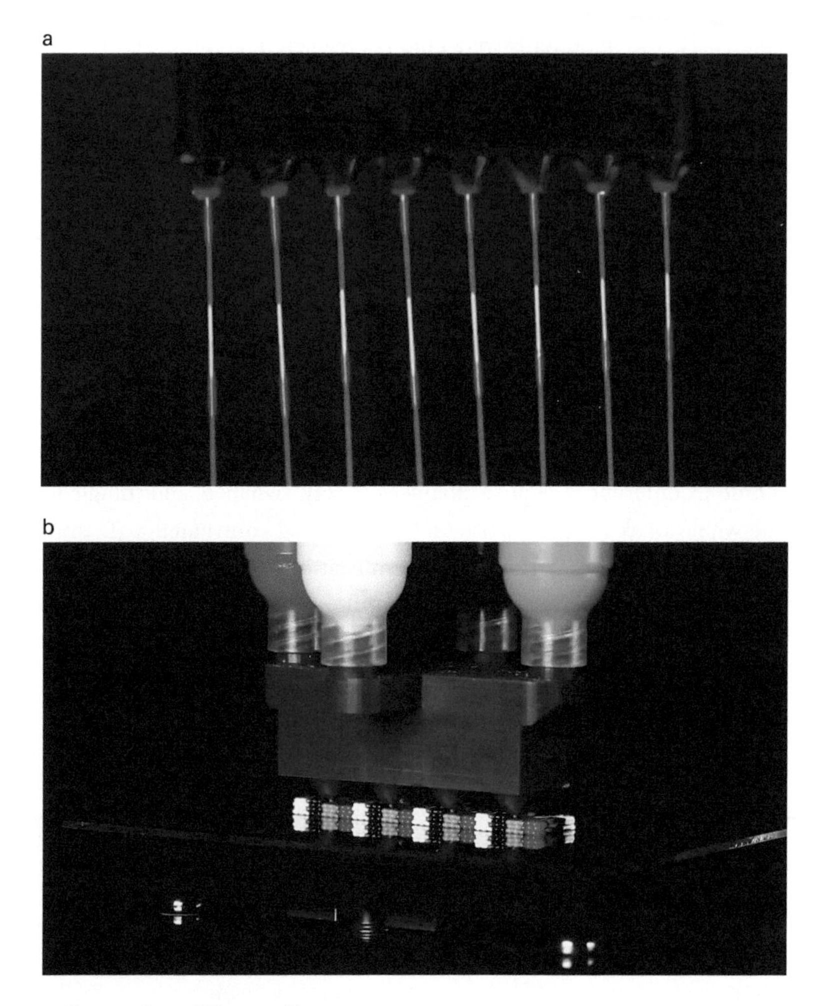

Figure 9.6a and Figure 9.6b
Multi-material 3D printing with multiple nozzles.
Images courtesy of Jennifer Lewis, Harvard University.

BOX 9.1: BIOGRAPHY
Jennifer Lewis

Jennifer Lewis was born in the midst of a hurricane, as if announcing the revolution she would cause with her new fluid dynamics approach to 3D printing. It was 1964, and her dad—an engineer—and her mom—an elementary school teacher—were living in Florida. As the hurricane approached, her parents took shelter in a Daytona Beach hospital, where Jennifer Ann Lewis was born. While Lewis grew up in many states, she went to high school in Palatine, a northwest suburb of Chicago. "I had several amazing teachers in mathematics and physics, which sparked my interest in science," she remembers. Fortunately, whenever they moved, her parents always chose their next home based on the quality of the school district. "My father encouraged me to pursue an engineering degree in college. Both my parents wanted me to go to college and saved money to ensure that I could," she says. Her parents are proud of her meteoric career. Lewis attended the University of Illinois at Urbana-Champaign (UIUC), a top-ranked engineering school, to study ceramic engineering. After graduating in 1986, she earned a PhD in ceramics science from the Department of Materials Science and Engineering with Michael Cima, who had just started his lab. Right after getting her degree, in December 1990, she joined the faculty of the Department of Materials Science and Engineering at UIUC.

gym and playing against her in a one-on-one basketball game—a passion that they both shared—Cesarano invited Lewis to spend her sabbatical at Sandia with his group in the fall of 1997. They co-advised a talented PhD student, James Smay, who did the seminal work on understanding and controlling the peculiar flow behavior of colloidal inks, whose viscosity changed with printing velocity.[17]

Smay's work helped catalyze Lewis's meteoric career. "As a materials scientist, I immediately recognized the potential of this technique for many applications," says Lewis. With this collaboration as a starting point, her lab expanded this platform to a broad set of functional, structural, and biological inks, new printhead designs, and other printing methods, pushing the boundaries of resolution and complexity.[18] While in lab talking with Smay, she made a serendipitous observation. She witnessed Smay translating the printhead that delivers the colloidal inks into a liquid reservoir to prevent drying between prints. "I noticed that the residual pressure within the nozzle led to the extrusion of an ink filament that maintained its shape within this liquid phase," Lewis recalls of this eureka moment. As she reflected on what she had seen and how surprising this observation was, "I realized that we could print structures within a reservoir," she adds. This discovery ushered in a new method, known as *embedded 3D printing*, in which inks are directly printed within a gel-like viscous liquid matrix. The function of the gel is to hold the material in place while the material hardens—afterward, the user removes the print from the vat.

Printing within a fluid environment had many other benefits. PhD student Greg Gratson used an alcohol-rich solution to drive the gelation of polyelectrolyte inks in a coagulation reservoir.[19] Polyelectrolytes are polymers that dissociate into ions in solution, acquiring the capacity to conduct electricity. Another talented PhD student, Willie Wu, came up with the idea of using the reservoir itself as a matrix. He patterned a "fugitive" ink (ink that is removed later) within a polymer–water matrix, which was the genesis for the group's work on vascularized human tissues.[20] In spring 2013, Lewis moved to the Wyss Institute at Harvard to explore other applications of her technology, such as biomedicine and robotics. At Harvard, Lewis collaborated with George Whitesides's and Robert Wood's groups to fabricate an octopus-inspired, entirely soft and autonomous robot nicknamed the Octobot (figure 9.7).[21] Powered by a chemical reaction controlled by PDMS microfluidics, the 3D-printed Octobot had no electronics. A reaction inside the Octobot transformed a small amount of hydrogen peroxide (acting as fuel) into a large volume of gas. This gas flowed then into the Octobot's arms, inflating them like a balloon. To produce the crawling motion, they inflated the arms in an alternating fashion by controlling the reaction with a microfluidic oscillator—a clever system of two PDMS microvalves that, connected to a constant input, diverts flow alternately to two different channels.[22] Soft robots have many potential applications, such as in the automated handling of fragile objects and in the remote exploration of delicate surfaces (e.g., remote surgery).

Figure 9.7
The Octobot, a 3D-printed soft robot that is self-powered via a chemical reaction in microfluidic channels. The arms of the Octobot span a ~5-centimeter-wide square and its head is about 2 centimeters at its widest point.
Image courtesy of Jennifer Lewis, Harvard University.

3D PRINTED ORGANS

Scientists and the public alike have hailed bioprinting of organs as one of the next frontiers of regenerative medicine. One of the challenges in bioprinting is that the biological structures that one can achieve with traditional 3D printers are discouragingly simplistic compared to the multicellular structures seen in human tissues, where many different cell types can coexist in a cubic millimeter within an extracellular matrix and the cells themselves often have complex three-dimensional shapes and orientations that are critical to their function (figure 9.8). Recently, two postdocs in Jennifer Lewis's lab, Mark Skylar-Scott and Sebastien Uzel, thought of a possible solution to this problem. They extended the embedded 3D printing concept "with an ingenious idea," as described by Lewis. Instead of filling the reservoir with a fluid, they filled it with living stem-cell-derived organoids. These are millimeter-sized spheres

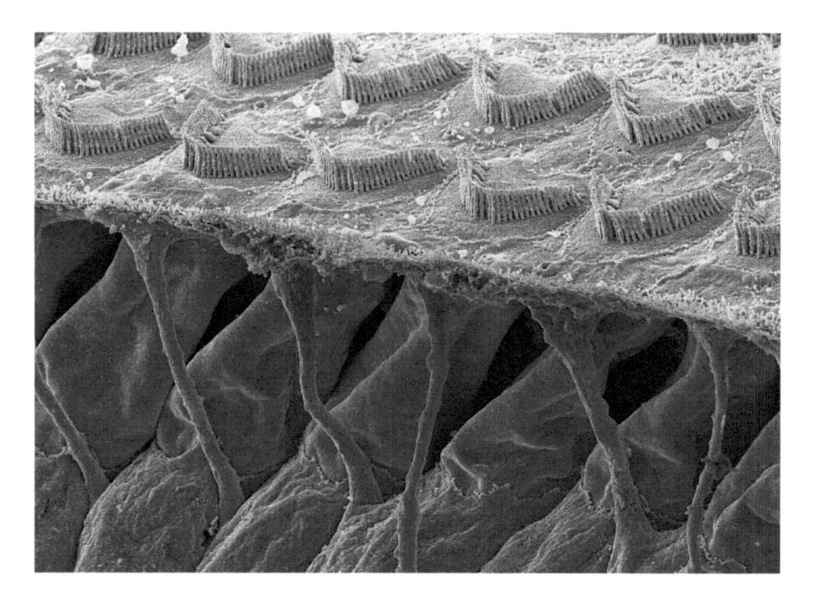

Figure 9.8

3D printers are not ready yet to fully reproduce the complexity of biological tissue. As an example of a structure that cannot be bioprinted with present technology, this scanning electron micrograph shows part of the inner ear from a guinea pig. The cilia on the top surface detect fluid movement in the inner ear when sound reaches the ear drum.
Image courtesy of Dr. David Furness.

of a type of human stem cells that are still in an embryo-like stage and can therefore grow and differentiate into any type of tissue. Organoids of many tissue types, containing multiple cell types, are possible. The ingeniousness resides in (correctly) assuming that the living organoid matrix would behave like a gel and allow them to print a 3D fugitive ink pattern within a functional tissue. After the organoids fused—they tend to do that—Skylar-Scott and Uzel removed the ink, leaving "vascular" conduits through which nutrients are perfused to sustain the solid tissue.[23] Because the tissue can be grown from the cells of a human donor, this technique could one day be used in clinics to build replacement for some tissues and organs.

Another challenge in bioprinting is the bioinks. How can the printed parts avoid rejection by the immune system? Tal Dvir's group at Tel Aviv University in Israel was able to genetically reprogram cells from the omentum (a large flat layer of fat tissue that surrounds every organ in the abdomen) to become heart muscle cells or endothelial (vascular) cells. Separately, they also decellularized human omentum tissue to concoct a hydrogel that could be used to make two types of bioinks, one containing heart muscle cells and the other one containing blood vessel–forming cells. Using Jennifer Lewis's embedded 3D printing approach, the Israeli researchers printed a heart and its vasculature by feeding the printer the two bioinks (figure 9.9).[24] By printing in immersion, the bioink avoids surface tension and drying artifacts that are common in other types of 3D printers. Although the heart is the size of a rabbit's and cannot pump fluids yet, the muscle cells did develop electrical contractility. "This is the first

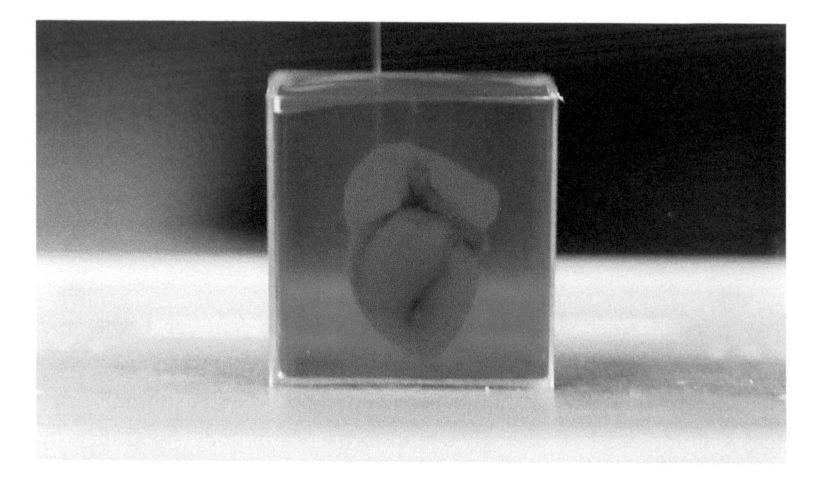

Figure 9.9
A miniature heart with its own vasculature in the process of being 3D printed with human cells using a bioprinter. The nozzle of the printer is visible as a thin needle half-submerged in a viscous fluid.
Image courtesy of Tal Dvir at Tel Aviv University.

time anyone anywhere has successfully engineered and printed an entire heart replete with cells, blood vessels, ventricles and chambers," Tal Dvir has said.[25] The 3D-printed bioinks and cells are not expected to cause an immune reaction because they are created from the patient's omentum.

Other 3D printing approaches may address the resolution limits of Lewis's approach. A form of high-resolution 3D printing called *stereolithography* that uses projected light patterns to print materials layer by layer (see section 3D-Printed Microfluidics in Chapter 5) might help, especially when used to print hydrogels. Hydrogels are polymers that, like Jello, contain large amounts of water and thus are very porous, resembling the consistency of biological tissue. This porosity is beneficial for fabrication because it allows researchers to thoroughly wash off all the excess chemicals that are needed for light exposure but would otherwise be harmful to cells. In 2019, Jordan Miller's group used stereolithography to 3D print stunning microfluidic architectures that seemed hitherto impossible to manufacture (figure 9.10).[26] The work was awarded the cover of *Science*. As a striking example, they fabricated artificial pulmonary "alveoli" in hydrogels that could expand during "inhalation" and exchange gases with a parallel blood vasculature network (figure 9.11). Such biomimetic tissue constructs are called to revolutionize the design of organs-on-chips and implantable devices.

* * *

The work of researchers from Andreas Manz to Jennifer Lewis, and many others, provides examples of how microfluidic devices have been shaping our everyday life. While behaving in odd ways at submillimeter scales, fluids can be extremely useful to us. Manz understood that, by miniaturizing the manipulation of fluids, chemical assays—such as the droplet and DNA chips used for the Human Genome Project—could be made faster, cheaper, and better. By introducing fluid automation at the nanoliter scale, Steve Quake ushered microfluidic experimentation into the modern era of big data and a Moore's law–like pace of progress. George Whitesides foresaw ways to free scientists of the complexities of

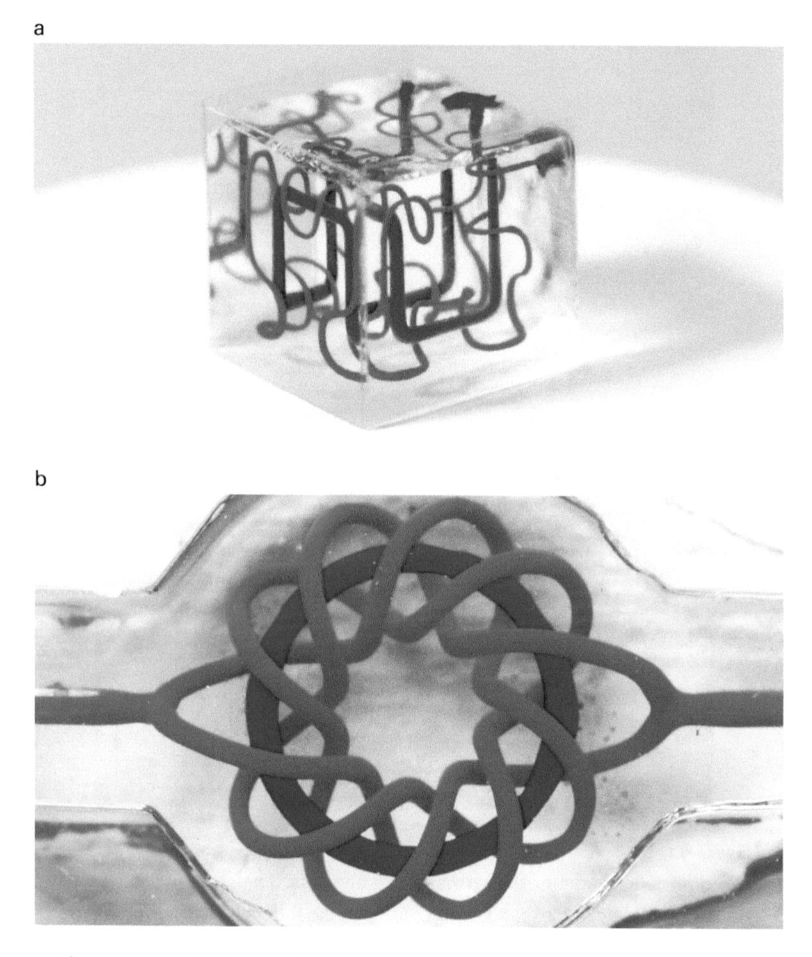

Figure 9.10a and Figure 9.10b
Complex 3D microfluidic channel architectures stereolithographically printed in a hydrogel lattice and filled with color dyes. The structures have been formed layer by layer using projected light and light-sensitive hydrogels. A (a) Hilbert cube and (b) torus are shown. The cube is 1.5 centimeters tall and the thinnest red tubes inside the torus are 1 millimeter in diameter.
Images courtesy of Jordan Miller, Rice University.

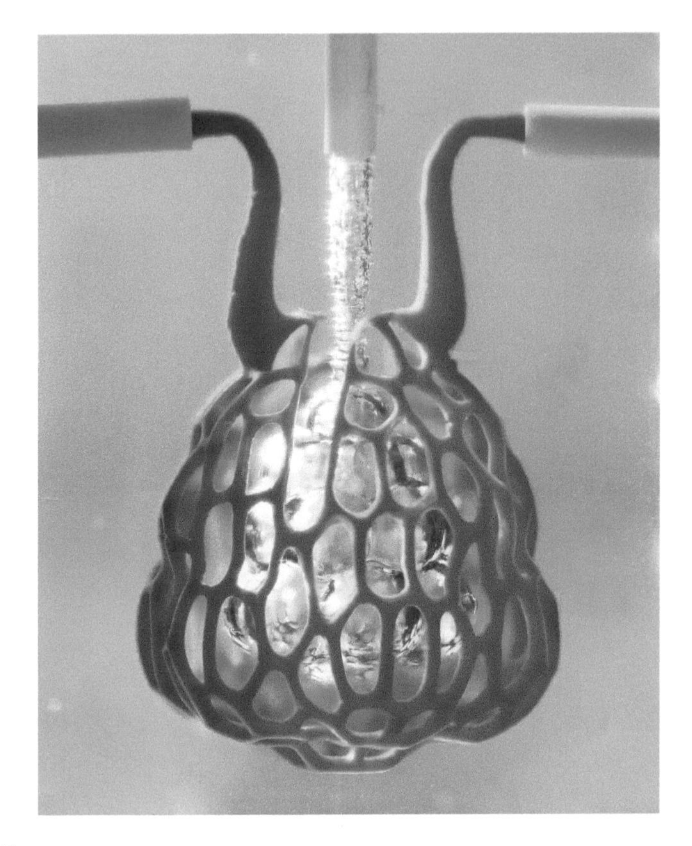

Figure 9.11

A biomimetic model of the human lung alveolus. The structure is about 5–6 millimeters wide and expands and contracts like the real analog. Deoxygenated red blood cells enter through one of the side inlets and circulate to become oxygenated through the branched network until they exit, fully oxygenated, through the other side. To mimic inhalation, the center inlet is pressurized with oxygen gas, which then diffuses through the hydrogel to the red blood cells, a process similar to how red blood cells in our lungs get reloaded with oxygen in the lung alveoli.

Image courtesy of Jordan Miller, Rice University.

manufacturing and operation of microfluidic devices, and by reducing them to their minimalistic expression in paper he was able to reach out to underserved communities. Dino Di Carlo and Mehmet Toner realized that particles and cells can enter a coordinated dance with fluids at high speeds, enabling devices that are capable of ultrafast, ultrasensitive processing of blood that one day might warn us of impending cancer. Lewis observed how viscosity changes with flow velocity and used it to devise a new class of 3D printing that has many applications, among them tissue printing. Many other, countless microfluidic engineers and scientists have contributed other stories that did not fit in this book—what I had time to tell you is just the tip of the iceberg of a burgeoning, extremely productive, alluring field. The success of microfluidic devices speaks to how essential fluids are, down to the smallest dimensions, for nature as well as for our technology-infused life.

NOTES

Prologue

1. Einstein, A. "Zur Elektrodynamik bewegter Körper" [On the Electrodynamics of Moving Bodies]. *Ann. Phys.* *17*, 891–921 (1905).

2. Galison, P. *Einstein's Clocks, Poincaré's Maps: Empires of Time*. Norton & Co., 2003.

3. Stern, M. John Quincy Adams argues to the US Supreme Court in the *Amistad* case that the natural state of mankind is freedom. *Curiously*. http://www.curiously.co/posts/john-quincy-adams-argues-to-the-us-supreme-court-in-the-amistad-case-that-the-natural-state-of-mankind-is-freedom (undated).

Introduction

1. Folch, A., Tejada, J., Peters, C. H. & Wrighton, M. S. Electron beam deposition of gold nanostructures in a reactive environment. *Appl. Phys. Lett.* *66*, 2080 (1995).

2. Folch, A., Wrighton, M. S. & Schmidt, M. A. Microfabrication of oxidation-sharpened silicon tips on silicon nitride cantilevers for atomic force microscopy. *J. Microelectromechanical Syst.* *6*, 303 (1997).

1. Tiny Channels Everywhere

1. Kim, Y.-C., Park, J.-H. & Prausnitz, M. R. Microneedles for drug and vaccine delivery. *Adv. Drug Deliv. Rev.* *64*, 1547–1568 (2012).

2. Kang, E., Jeong, G. S., Choi, Y. Y., Lee, K. H., Khademhosseini, A. & Lee, S. H. Digitally tunable physicochemical coding of material composition and topography in continuous microfibres. *Nat. Mater.* *10*, 877–883 (2011).

3. Folch, A. *Introduction to BioMEMS*. CRC Press, 2012.

4. Hauksbee, F. An experiment made at Gresham-College, shewing that the seemingly spontaneous ascention of water in small tubes open at both ends is the same in vacuo as in the open air. *Philos. Trans.* 25, 2223–2224 (1706).

5. Jurin, J. An account of some experiments shown before the Royal Society; with an enquiry into the cause of the ascent and suspension of water in capillary tubes. *Philos. Trans.* 30, 739–747 (1718).

6. Young, T. An essay on the cohesion of fluids. *Philos. Trans. R. Soc. London* 95, 65–87 (1805); Laplace, P. S. Supplément au dixième livre du Traité de Mécanique Céleste. In *Traité de Mécanique Céleste*, vol. 4, pp. 1–79 (1805).

7. Nakayama, Y. *Introduction to Fluid Mechanics*. Butterworth-Heinemann, 1998.

8. *See* Nakayama (1998).

9. Graham, T. On the law of diffusion of gases. *Philos. Mag.* 2, 175–358 (1833).

10. Philpot, J. St. L. The use of thin layers in electrophoretic separation. *Trans. Faraday Soc.* 35, 38–46 (1940).

11. Haugaard, G. & Kroner, T. D. Partition chromatography of amino acids with applied voltage. *J. Am. Chem. Soc.* 70, 2135–2137 (1948).

12. Grassman, W. & Hannig, K. A simple process for the continuous separation of substance mixtures on filter paper by electrophoresis. *Naturwissenschaften* 37, 397 (1950).

13. Jorgenson, J. W. & Lukacs, K. D. A. Zone electrophoresis in open-tubular glass capillaries. *Anal. Chem.* 53, 1298–1302 (1981); Jorgenson, J. & Lukacs, K. Capillary zone electrophoresis. *Science* 222, 266–272 (1983).

14. NASA. NASA's electrophoresis chip. https://ntrs.nasa.gov/api/citations/19880001596/downloads/19880001596.pdf (1987).

15. Petersen, K. E. Silicon as a mechanical material. *Proc. IEEE* 70, 420–457 (1982).

2. The Power of Droplets

1. Elmqvist, R. Measuring instrument of the recording type. US Patent 2,566,443, filed September 4, 1949, and issued September 4, 1951.

2. Sweet, R. G. High frequency recording with electrostatically deflected ink jets. *Rev. Sci. Instrum.* 36, 131–136 (1965).

3. Zoltan, S. I. Pulsed droplet ejecting system. US Patent 3,857,049, filed October 12, 1973, and issued December 24, 1974.

4. Kyser, E. L. & Sears, S. B. Method and apparatus for recording with writing fluids and drop projection means therefor. US Patent 3,946,398, filed June 29, 1970, and issued March 23, 1976.

5. Nielsen, N. J. History of ThinkJet printhead development. *Hewlett-Packard J. 36*, 4–10 (1985). http://www.hpl.hp.com/hpjournal/pdfs/IssuePDFs/1985-05.pdf.

6. *See* Nielsen (1985).

7. Samaun, S., Wise, K. D., Nielsen, E. D. & Angell, J. B. An IC piezoresistive pressure sensor for biomedical instrumentation. *IEEE Trans. Biomed. Eng. BME 20*, 101–109 (1973).

8. Rafkin, L. The founder. *Forbes.* https://www.forbes.com/asap/2001/0402/MEMS_xtra1_print.html (2001).

9. Wise, K. D., Angell, J. B. & Starr, A. An integrated-circuit approach to extracellular microelectrodes. *IEEE Trans. Biomed. Eng.* 17, 238–247 (1970).

10. Butrica, A. J. NASA's role in the development of MEMS. Chapter 4 of *Historical Studies in the Societal Impact of Spaceflight*. NASA SP-2015-4803, 251–330. https://www.nasa.gov/sites/default/files/atoms/files/historical-studies-societal-impact-spaceflight-ebook_tagged.pdf (2015).

11. *See* Samaun (1973).

12. Terry, S. C., Jerman, J. H. & Angell, J. B. A gas chromatographic air analyzer fabricated on a silicon wafer. *IEEE Trans. Electron Devices 26*, 1880–1886 (1979).

13. Angell, J. B., Terry, S. C. & Barth, P. W. Silicon micromechanical devices. *Sci. Am.* https://www.scientificamerican.com/article/silicon-micromechanical-devices (1983).

14. Petersen, K. E. Silicon as a mechanical material. *Proc. IEEE 70*, 420–457 (1982).

15. *See* Samaun (1973).

16. Bassous, E., Taub, H. H. & Kuhn, L. Ink jet printing nozzle arrays etched in silicon. *Appl. Phys. Lett. 31*, 135–137 (1977); Kuhn, L., Bassous, E. & Lane, R. Silicon charge electrode array for ink jet printing. *IEEE Trans. Electron Devices* 25, 1257–1260 (1978).

17. Waggener, H. A., Kragness, R. C. & Taylor, A. L. Anisotropic etching for forming isolation slots. *Electronics* 40, 274–276 (1967).

18. Bassous, E., Taub, H. H. & Kuhn, L. Ink jet printing nozzle arrays etched in silicon. *Appl. Phys. Lett. 31*, 135–137 (1977).

19. Le, H. P. Progress and trends in ink-jet printing technology. *J. Imaging Sci. Technol. 42*, 49–62 (1998).

20. Alfaro, D. What is an emulsion in the culinary arts? *The Spruce Eats*. https://www.thespruceeats.com/what-is-an-emulsion-995655 (updated August 16, 2019).

21. *See* Alfaro (2019).

22. *See* Alfaro (2019).

23. Kung, C. Y., Barnes, M. D., Lermer, N., Whitten, W. B. & Ramsey, J. M. Confinement and manipulation of individual molecules in attoliter volumes. *Anal. Chem. 70*, 658–661 (1998); Tawfik, D. S. & Griffiths, A. D. Man-made cell-like compartments for molecular evolution. *Nat. Biotechnol. 16*, 652–656 (1998); Chiu, D. T., Wilson, C. F., Ryttsen, F., Stromberg, A., Farre, C., Karlsson, A., Nordholm, S. et al. Chemical transformations in individual ultrasmall biomimetic containers. *Science 283*, 1892–1895 (1999).

24. Kawakatsu, T., Kikuchi, Y. & Nakajima, M. Regular-sized cell creation in microchannel emulsification by visual microprocessing method. *J. Am. Oil Chem. Soc. 74*, 317–321 (1997).

25. Thorsen, T., Roberts, R. W., Arnold, F. H. & Quake, S. R. Dynamic pattern formation in a vesicle-generating microfluidic device. *Phys. Rev. Lett. 86*, 4163–4166 (2001).

26. Anna, S. L., Bontoux, N. & Stone, H. A. Formation of dispersions using "flow focusing" in microchannels. *Appl. Phys. Lett. 82*, 364–366 (2003).

27. Teh, S., Lin, R., Hung, L. & Lee, A. P. Droplet microfluidics. *Lab Chip 8*, 198–220 (2008).

28. *See* Teh (2008).

29. Li, L., Mustafi, D., Fu, Q., Tereshko, V., Chen, D. L., Tice, J. D. & Ismagilov, R. F. Nanoliter microfluidic hybrid method for simultaneous screening and optimization validated with crystallization of membrane proteins. *Proc. Natl. Acad. Sci. U. S. A. 103*, 19243–19248 (2006).

30. Agresti, J. J., Antipov, E., Abate, A. R., Ahn, K., Rowat, A. C., Baret, J.-C., Marquez, M., Klibanov, A. M., Griffiths, A. D. & Weitz, D. A. Ultrahigh-throughput

screening in drop-based microfluidics for directed evolution. *Proc. Natl. Acad. Sci. U. S. A. 107*, 4004–4009 (2010).

31. Macosko, E. Z., Basu, A., Satija, R., Nemesh, J., Shekhar, K., Goldman, M., Tirosh, I. et al. Highly parallel genome-wide expression profiling of individual cells using nanoliter droplets. *Cell 161*, 1202–1214 (2015).

32. Lan, F., Demaree, B., Ahmed, N. & Abate, A. R. Single-cell genome sequencing at ultra-high-throughput with microfluidic droplet barcoding. *Nat. Biotechnol. 35*, 640–646 (2017).

33. Edel, J. B., Fortt, R., deMello, J. C. & deMello, A. J. Microfluidic routes to the controlled production of nanoparticles. *Chem. Commun. 2*, 1136–1137 (2002).

34. Niu, X., Gielen, F., Edel, J. B. & deMello, A. J. A microdroplet dilutor for high-throughput screening. *Nat. Chem. 3*, 437–442 (2011).

35. Beans, C. Science and culture: Universities move science labs to the kitchen. *Proc. Natl. Acad. Sci. U. S. A. 117*, 20982–20985 (2020).

36. Black, J. Foam 101? Chefs Andrés, Adrià will teach at Harvard. *Washington Post*, March 24 (2010).

37. Sweeney, S. In good taste. *The Harvard Gazette*, September 8 (2010). https://news.harvard.edu/gazette/story/2010/09/in-good-taste/.

38. *See* Black (2010).

39. Clausell-Tormos, J., Lieber, D., Baret, J.-C., El-Harrak, A., Miller, O. J., Frenz, L., Blouwolff, J. et al. Droplet-based microfluidic platforms for the encapsulation and screening of mammalian cells and multicellular organisms. *Chem. Biol. 15*, 427–437 (2008); Köster, S., Angilè, F. E., Duan, H., Agresti, J. J., Wintner, A., Schmitz, C., Rowat, A. C. et al. Drop-based microfluidic devices for encapsulation of single cells. *Lab Chip 8*, 1110 (2008).

40. Adrià, F, Soler, J. & Adrià, A. *El Bulli, 1983–1993*, vol. 1. RBA Practica, 2004.

41. ChefSteps. The science of spherification. https://www.chefsteps.com/activities/the-science-of-spherification.

42. Lim, F. & Sun, A. M. Microencapsulated islets as bioartificial endocrine pancreas. *Science 210*, 908–910 (1980); De Vos, P., Van Hoogmoed, C. G., Van Zanten, J., Netter, S., Strubbe, J. H. & Busscher, H. J. Long-term biocompatibility, chemistry, and function of microencapsulated pancreatic islets. *Biomaterials 24*, 305–312 (2003).

43. Read, T. A., Sorensen, D. R., Mahesparan, R., Enger, P., Timpl, R., Olsen, B. R., Hjelstuen, M. H. B., Haraldseth, O. & Bjerkvig, R. Local endostatin treatment of gliomas administered by microencapsulated producer cells. *Nat. Biotechnol. 19*, 29–34 (2001).

44. Sugiura, S., Oda, T., Izumida, Y., Aoyagi, Y., Satake, M., Ochiai, A., Ohkohchi, N. & Nakajima, M. Size control of calcium alginate beads containing living cells using micro-nozzle array. *Biomaterials 26*, 3327–3331 (2005).

45. Tan, W. H. & Takeuchi, S. Monodisperse alginate hydrogel microbeads for cell encapsulation. *Adv. Mater. 19*, 2696–2701 (2007); Mazutis, L., Vasiliauskas, R. & Weitz, D. A. Microfluidic production of alginate hydrogel particles for antibody encapsulation and release. *Macromol. Biosci. 15*, 1641–1646 (2015); Utech, S., Prodanovic, R., Mao, A. S., Ostafe, R., Mooney, D. J. & Weitz, D. A. Microfluidic generation of monodisperse, structurally homogeneous alginate microgels for cell encapsulation and 3D cell culture. *Adv. Healthc. Mater. 4*, 1628–1633 (2015); Chen, Q., Utech, S., Chen, D., Prodanovic, R., Lin, J. M. & Weitz, D. A. Controlled assembly of heterotypic cells in a core-shell scaffold: Organ in a droplet. *Lab Chip 16*, 1346–1349 (2016).

46. Chen, L. H., Cheng, L. C. & Doyle, P. S. Nanoemulsion-loaded capsules for controlled delivery of lipophilic active ingredients. *Adv. Sci. 7*, 2001677 (2020).

Chapter 3

1. van der Schoot, B. H. & Bergveld, P. An ISFET-based microliter titrator integration of a chemical sensor-actuator system. *Sens. Actuators 8*, 11–22 (1985); van der Schoot, B. H. & Bergveld, P. The pH-static enzyme sensor. An ISFET-based enzyme sensor, insensitive to the buffer capacity of the sample. *Anal. Chim. Acta 199*, 157–160 (1987).

2. Manz, A., Miyahara, Y., Miura, J., Watanabe, Y., Miyagi, H. & Sato, K. Design of an open-tubular column liquid chromatograph using silicon chip technology. *Sens. Actuators B 1*, 249–255 (1990).

3. Widmer, H. M. Trends in industrial analytical chemistry. *Trends Anal. Chem. 2*, 1–3 (1983).

4. Manz, A. Manz speaker deck at Ciba Geigy. https://speakerdeck.com/amzamz/miniaturized-total-chemical-analysis-systems-a-novel-concept-for-chemical-sensing (1989).

5. Bloomstein, T. M. & Ehrlich, D. J. Laser-chemical three-dimensional writing for microelectromechanics and application to standard-cell microfluidics. *J. Vac. Sci. Technol. B 10*, 2671 (1992).

6. Manz, A., Harrison, D. J., Verpoorte, E. M. J., Fettinger, J. C., Paulus, A., Lüdi, H. & Widmer, H. M. Planar chips technology for miniaturization and integration of separation techniques into monitoring systems. Capillary electrophoresis on a chip. *J. Chromatogr. A* 593, 253–258 (1992).

7. Harrison, D. J. A personal stroll through the historical development of Canadian microfluidics. *Lab Chip 13*, 2500–2503 (2013).

8. Shoji, S., Esashi, M. & Matsuo, T. Prototype miniature blood gas analyser fabricated on a silicon wafer. *Sens. Actuators 14*, 101–107 (1988).

9. van Lintel, H. T. G., van De Pol, F. C. M. & Bouwstra, S. A piezoelectric micropump based on micromachining of silicon. *Sens. Actuators A 15*, 153–167 (1988).

10. Smits, J. G. Piezoelectric micropump with three valves working peristaltically. *Sens. Actuators A 21*, 203–206 (1990).

11. Harrison, D. J. A personal stroll through the historical development of Canadian microfluidics. *Lab Chip 13*, 2500–2503 (2013).

12. Harrison, D. J., Manz, A., Fan, Z. H., Ludi, H. & Widmer, H. M. Capillary electrophoresis and sample injection systems integrated on a planar glass chip. *Anal. Chem.* 64, 1926 (1992); Manz, A., Harrison, D. J., Verpoorte, E. M. J., Fettinger, J. C., Paulus, A., Ludi, H. & Widmer, H. M. Planar chips technology for miniaturization and integration of separation techniques into monitoring systems—Capillary electrophoresis on a chip. *J. Chromatogr.* 593, 253 (1992); Harrison, D. J., Fluri, K., Seiler, K., Fan, Z. H., Effenhauser, C. S. & Manz, A. Micromachining a miniaturized capillary electrophoresis-based chemical-analysis system on a chip. *Science 261*, 895 (1993); Effenhauser, C. S., Manz, A. & Widmer, H. M. Glass chips for high-speed capillary electrophoresis separations with submicrometer plate heights. *Anal. Chem.* 65, 2637 (1993).

13. Harrison, D. J., Manz, A., Fan, Z. H., Ludi, H. & Widmer, H. M. Capillary electrophoresis and sample injection systems integrated on a planar glass chip. *Anal. Chem.* 64, 1926 (1992).

14. Manz, A., Graber, N. & Widmer, H. M. Miniaturized total chemical analysis systems: A novel concept for chemical sensing. *Sens. Actuators B 1*, 244–248 (1990); Manz, A., Harrison, D. J., Verpoorte, E. M. J., Fettinger, J. C., Lüdi, H. & Widmer,

H. M. Miniaturization of chemical-analysis systems—A look into next century technology or just a fashionable craze. *Trac-Trend Anal. Chem. 10*, 144 (1991); Harrison, D. J., Glavina, P. G. & Manz, A. Towards miniaturized electrophoresis and chemical-analysis systems on silicon—An alternative to chemical sensors. *Sens. Actuator B 10*, 107 (1993); Kopp, M. U., Crabtree, H. J. & Manz, A. Developments in technology and applications of microsystems. *Curr. Opin. Chem. Biol. 1*, 410 (1997).

15. Manz, A., Graber, N. & Widmer, H. M. Miniaturized total chemical analysis systems: A novel concept for chemical sensing. *Sens. Actuators B. 1*, 244–248 (1990).

16. Harrison, D. J., Fluri, K., Seiler, K., Fan, Z. H., Effenhauser, C. S. & Manz, A. Micromachining a miniaturized capillary electrophoresis-based chemical-analysis system on a chip. *Science 261*, 895 (1993).

17. Little, W. A. Microminiature refrigeration. *Rev. Sci. Instrum.* 55, 661–680 (1984); Wu, P. & Little, W. A. Measurement of the heat transfer characteristics of gas flow in fine channel heat exchangers used for microminiature refrigerators. *Cryogenics (Guildf). 24*, 415–420 (1984).

18. van Erp, R., Soleimanzadeh, R., Nela, L., Kampitsis, G. & Matioli, E. Co-designing electronics with microfluidics for more sustainable cooling. *Nature 585*, 211–216 (2020).

19. Huang, X. C., Quesada, M. A. & Mathies, R. A. DNA sequencing using capillary array electrophoresis. *Anal. Chem. 64*, 2149–2154 (1992); Huang, X. C., Quesada, M. A. & Mathies, R. A. Capillary array electrophoresis using laser-excited confocal fluorescence detection. *Anal. Chem. 64*, 967–972 (1992); Mathies, R. A. & Huang, X. C. Capillary array electrophoresis: an approach to high-speed, high-throughput DNA sequencing. *Nature 359*, 167–169 (1992).

20. Woolley, A. T., Sensabaugh, G. F. & Mathies, R. A. High-speed DNA genotyping using microfabricated capillary array electrophoresis chips. *Anal. Chem. 69*, 2181–2186 (1997).

21. Woolley, A. T. & Mathies, R. A. Ultra-high-speed DNA sequencing using capillary electrophoresis chips. *Anal. Chem. 67*, 3676–3680 (1995).

22. Woolley, Sensabaugh & Mathies, R. A. High-speed DNA genotyping.

23. Woolley, A. T., Hadley, D., Landre, P., DeMello, A. J., Mathies, R. A. & Northrup, M. A. Functional integration of PCR amplification and capillary electrophoresis in a microfabricated DNA analysis device. *Anal. Chem. 68*, 4081–4086 (1996).

24. Kopp, M. U., De Mello, A. J. & Manz, A. Chemical amplification: Continuous-flow PCR on a chip. *Science 280*, 1046–1048 (1998).

25. Northrup, M. A., Ching, M. T., White, R. M. & Watson, R. T. DNA amplification in a microfabricated reaction chamber. In *Proc., 7th International Conference on Solid-State Sensors and Actuators (Transducers '93), Yokohama, Japan (7–10 June)*, p. 924 (1993).

26. Woolley, A. T., Hadley, D., Landre, P., DeMello, A. J., Mathies, R. A. & Northrup, M. A. Functional integration of PCR amplification and capillary electrophoresis in a microfabricated DNA analysis device. *Anal. Chem. 68*, 4081–4086 (1996); Wilding, P., Shoffner, M. A. & Kricka, L. J. PCR in a silicon microstructure. *Clin. Chem. 40*, 1815–1818 (1994); Northrup, M. A., Gonzalez, C., Hadley, D., Hills, R. F., Landre, P., Lehew, S., Saw, R., Sninsky, J. J. & Watson, R. A MEMS-based miniature DNA analysis system. In *Proc., International Solid-State Sensors Actuators Conference (Transducers '95)*, vol. 1, pp. 764–767 (1995); Burns, M. A., Mastrangelo, C. H., Sammarco, T. S., Man, F. P., Webster, J. R., Johnsons, B. N., Foerster, B. et al. Microfabricated structures for integrated DNA analysis. *Proc. Natl. Acad. Sci. U. S. A. 93*, 5556–5561 (1996); Burns, M. A., Johnson, B. N., Brahmasandra, S. N., Handique, K., Webster, J. R., Krishnan, M., Sammarco, T. S. et al. An integrated nanoliter DNA analysis device. *Science 282*, 484–487 (1998).

27. Woolley, A. T., Hadley, D., Landre, P., DeMello, A. J., Mathies, R. A. & Northrup, M. A. Functional integration of PCR amplification and capillary electrophoresis in a microfabricated DNA analysis device. *Anal. Chem. 68*, 4081–4086 (1996).

28. Lagally, E. T., Emrich, C. A. & Mathies, R. A. Fully integrated PCR-capillary electrophoresis microsystem for DNA analysis. *Lab Chip 1*, 102 (2001).

29. Lagally, E. T., Medintz, I. & Mathies, R. A. Single-molecule DNA amplification and analysis in an integrated microfluidic device. *Anal. Chem. 73*, 565–570 (2001).

30. Simpson, P. C., Roach, D., Woolley, A. T., Thorsen, T., Johnston, R., Sensabaugh, G. F. & Mathies, R. A. High-throughput genetic analysis using microfabricated 96-sample capillary array electrophoresis microplates. *Proc. Natl. Acad. Sci. U. S. A. 95*, 2256–2261 (1998); Shi, Y., Simpson, P. C., Scherer, J. R., Wexler, D., Skibola, C., Smith, M. T. & Mathies, R. A. Radial capillary array electrophoresis microplate and scanner for high-performance nucleic acid analysis. *Anal. Chem. 71*, 5354–5361 (1999).

31. Emrich, C. A., Tian, H., Medintz, I. L. & Mathies, R. A. Microfabricated 384-lane capillary array electrophoresis bioanalyzer for ultrahigh-throughput genetic analysis. *Anal. Chem.* 74, 5076–5083 (2002).

32. Breslow, R. The origin of homochirality in amino acids and sugars on prebiotic earth. *Tetrahedron Lett.* 52, 4228–4232 (2011).

33. Hutt, L. D., Glavin, D. P., Bada, J. L. & Mathies, R. A. Microfabricated capillary electrophoresis amino acid chirality analyzer for extraterrestrial exploration. *Anal. Chem.* 71, 4000–4006 (1999).

34. Skelley, A. M. & Mathies, R. A. Chiral separation of fluorescamine-labeled amino acids using microfabricated capillary electrophoresis devices for extraterrestrial exploration. *J. Chromatogr. A 1021,* 191–199 (2003).

35. Skelley, A. M., Scherer, J. R., Aubrey, A. D., Grover, W. H., Ivester, R. H. C., Ehrenfreund, P., Grunthaner, F. J., Bada, J. L. & Mathies, R. A. Development and evaluation of a microdevice for amino acid biomarker detection and analysis on Mars. *Proc. Natl. Acad. Sci. U. S. A. 102,* 1041–1046 (2005).

36. Skelley, A. M., Scherer, J. R., Aubrey, A. D., Grover, W. H., Ivester, R. H. C., Ehrenfreund, P., Grunthaner, F. J., Bada, J. L. & Mathies, R. A. Development and evaluation of a microdevice for amino acid biomarker detection and analysis on Mars. *Proc. Natl. Acad. Sci. 102,* 1041–1046 (2005); Skelley, A. M., Aubrey, A. D., Willis, P. A., Amashukeli, X., Ehrenfreund, P., Bada, J. L., Grunthaner, F. J. & Mathies, R. A. Organic amine biomarker detection in the Yungay region of the Atacama Desert with the Urey instrument. *J. Geophys. Res. Biogeosciences 112,* 1–10 (2007).

37. Jensen, E. C., Grover, W. H. & Mathies, R. A. Micropneumatic digital logic structures for integrated microdevice computation and control. *J. Microelectromechanical Syst. 16,* 1378–1385 (2007).

38. Rhee, M. & Burns, M. A. Microfluidic pneumatic logic circuits and digital pneumatic microprocessors for integrated microfluidic systems. *Lab Chip* 9, 3131–3143 (2009).

39. Mosadegh, B., Kuo, C.-H., Tung, Y.-C., Torisawa, Y., Bersano-Begey, T., Tavana, H. & Takayama, S. Integrated elastomeric components for autonomous regulation of sequential and oscillatory flow switching in microfluidic devices. *Nat. Phys. 6,* 433–437 (2010).

40. Duncan, P. N., Nguyen, T. V. & Hui, E. E. Pneumatic oscillator circuits for timing and control of integrated microfluidics. *Proc. Natl. Acad. Sci. U. S. A. 110,* 18104–18109 (2013).

41. Wehner, M., Truby, R. L., Fitzgerald, D. J., Mosadegh, B., Whitesides, G. M., Lewis, J. A. & Wood, R. J. An integrated design and fabrication strategy for entirely soft, autonomous robots. *Nature 536*, 451–455 (2016).

Chapter 4

1. Rajendran, R. & Rayman, G. Point-of-care blood glucose testing for diabetes care in hospitalized patients: An evidence-based review. *J. Diabetes Sci. Technol. 8*, 1081–1090 (2014).

2. Moodley, N., Ngxamngxa, U., Turzyniecka, M. J. & Pillay, T. S. Historical perspectives in clinical pathology: A history of glucose measurement. *J. Clin. Pathol. 68*, 258–264 (2015).

3. Maumené, E. J. Sur un nouveau réactif pour distinguer la présence du sucre dans certains liquides. *J. Pharm. 17*, 368–370 (1850).

4. Oliver, G. On bedside urinary tests: Detection of sugar in the urine by means of test papers. *Lancet 121*, 858–860 (1883).

5. *See* Moodley (2015).

6. Free, A. H., Adams, E. C., Kercher, M. L., Free, H. M. & Cook, M. H. Simple specific test for urine glucose. *Clin. Chem. 3*, 163–167 (1957).

7. Kohn, J. A rapid method of estimating blood-glucose ranges. *Lancet 273*, 119–121 (1957).

8. *See* Moodley (2015).

9. Updike, S. J. & Hicks, G. P. Reagentless substrate analysis with immobilized enzymes. *Science 158*, 270–272 (1967); Updike, S. J. & Hicks, G. P. The enzyme electrode. *Nature 214*, 986 (1967).

10. *See* Moodley (2015).

11. Bernstein, R. K. Blood glucose control. *Arch Intern Med. 141*, 267–268 (1981).

12. *See* Bernstein (1981).

13. *See* Moodley (2015).

14. Heller, A. Electrical wiring of redox enzymes. *Acc. Chem. Res. 23*, 128–134 (1990); Heller, A. Electrical connection of enzyme redox centers to electrodes. *J. Phys. Chem. 96*, 3579–3587 (1992); Ohara, T. J., Vreeke, M. S., Battaglini, F. & Heller, A.

Bienzyme sensors based on electrically wired peroxidase. *Electroanalysis* 5, 825–831 (1993).

15. Heller, A. Implanted electrochemical glucose sensors for the management of diabetes. *Annu. Rev. Biomed. Eng.* 1, 153–175 (1999).

16. Feldman, B., McGarraugh, G., Heller, A., Bohannon, N., Skyler, J., DeLeeuw, E. & Clarke, D. Freestyle(TM): A small-volume electrochemical glucose sensor for home blood glucose testing. *Diabetes Technol. Ther.* 2, 221–229 (2000).

17. Lauks, I. R. Microfabricated biosensors and microanalytical systems for blood analysis. *Acc. Chem. Res.* 31, 317–324 (1998).

18. Bagnall, J. Success is in the blood—The toil and sweat of biotech pioneer Imants Lauks. *Ottawa Citizen* (2010).

19. Syed, D. 2011 award presented to Imant Lauks. *The Curious Chemist* (2012).

20. *See* Bagnall (2010).

21. Lauks, Microfabricated biosensors and microanalytical systems.

22. Calem, R. E. Moving the common blood test closer to the patient. *New York Times*. https://www.nytimes.com/1992/06/21/business/technology-moving-the-common-blood-test-closer-to-the-patient.html (1992).

23. Stotler, B. A. & Kratz, A. Analytical and clinical performance of the epoc blood analysis system: Experience at a large tertiary academic medical center. *Am. J. Clin. Pathol.* 140, 715–720 (2013).

24. Metz, S., Holzer, R. & Renaud, P. Polyimide-based microfluidic devices. *Lab Chip* 1, 29–34 (2001).

25. Roberts, M. A., Rossier, J. S., Bercier, P. & Girault, H. UV laser machined polymer substrates for the development of microdiagnostic systems. *Anal. Chem.* 69, 2035–2042 (1997); Rossier, J. S., Roberts, M. A., Ferrigno, R. & Girault, H. H. Electrochemical detection in polymer microchannels. *Anal. Chem.* 71, 4294–4299 (1999); Rossier, J. S., Schwarz, A., Reymond, F., Ferrigno, R., Bianchi, F. & Girault, H. H. Microchannel networks for electrophoretic separations. *Electrophoresis* 20, 727–731 (1999).

26. Ruano-López, J. M., Agirregabiria, M., Olabarria, G., Verdoy, D., Bang, D. D., Bu, M., Wolff, A. et al. The SmartBioPhone™, a point of care vision under development through two European projects: OPTOLABCARD and LABONFOIL. *Lab*

Chip 9, 1495 (2009); Focke, M., Kosse, D., Müller, C., Reinecke, H., Zengerle, R. & von Stetten, F. Lab-on-a-Foil: Microfluidics on thin and flexible films. *Lab Chip* 10, 1365 (2010).

27. Roberts, M. A., Rossier, J. S., Bercier, P. & Girault, H. UV laser machined polymer substrates for the development of microdiagnostic systems. *Anal. Chem.* 69, 2035–2042 (1997).

28. McCormick, R. M., Nelson, R. J., Alonso-Amigo, M. G., Benvegnu, D. J. & Hooper, H. H. Microchannel electrophoretic separations of DNA in injection-molded plastic substrates. *Anal. Chem.* 69, 2626–2630 (1997).

29. Martynova, L., Locascio, L. E., Gaitan, M., Kramer, G. W., Christensen, R. G. & MacCrehan, W. A. Fabrication of plastic microfluid channels by imprinting methods. *Anal. Chem.* 69, 4783–4789 (1997).

30. *See* McCormick (1997).

31. Dang, F., Tabata, O., Kurokawa, M., Ewis, A. A., Zhang, L., Yamaoka, Y., Shinohara, S., Shinohara, Y., Ishikawa, M. & Baba, Y. High-performance genetic analysis on microfabricated capillary array electrophoresis plastic chips fabricated by injection molding. *Anal. Chem.* 77, 2140–2146 (2005).

32. Harrison, D. J., Manz, A., Fan, Z. H., Ludi, H. & Widmer, H. M. Capillary electrophoresis and sample injection systems integrated on a planar glass chip. *Anal. Chem.* 64, 1926 (1992).

33. Sia, S. K., Linder, V., Parviz, B. A., Siegel, A. & Whitesides, G. M. An integrated approach to a portable and low-cost immunoassay for resource-poor settings. *Angew. Chemie-International Ed.* 43, 498–502 (2004); Linder, V., Sia, S. K. & Whitesides, G. M. Reagent-loaded cartridges for valveless and automated fluid delivery in microfluidic devices. *Anal. Chem.* 77, 64–71 (2005).

34. Chin, C. D., Laksanasopin, T., Cheung, Y. K., Steinmiller, D., Linder, V., Parsa, H., Wang, J. et al. Microfluidics-based diagnostics of infectious diseases in the developing world. *Nat. Med.* 17, 1015–1019 (2011).

35. Becker, H. & Heim, U. Hot embossing as a method for the fabrication of polymer high aspect ratio structures. *Sens. Actuators, A* 83, 130–135 (2000); Wabuyele, M. B., Ford, S. M., Stryjewski, W., Barrow, J. & Soper, S. A. Single molecule detection of double-stranded DNA in poly(methylmethacrylate) and polycarbonate microfluidic devices. *Electrophoresis* 22, 3939–3948 (2001); Kricka, L. J., Fortina, P., Panaro, N. J., Wilding, P., Alonso-Amigo, G. & Becker, H. Fabrication of plastic

microchips by hot embossing. *Lab Chip* 2, 1–4 (2002); Lee, G. Bin, Chen, S. H., Huang, G. R., Sung, W. C. & Lin, Y. H. Microfabricated plastic chips by hot embossing methods and their applications for DNA separation and detection. *Sens. Actuators, B* 75, 142–148 (2001).

36. Martynova, L., Locascio, L. E., Gaitan, M., Kramer, G. W., Christensen, R. G. & MacCrehan, W. A. Fabrication of plastic microfluid channels by imprinting methods. *Anal. Chem.* 69, 4783–4789 (1997).

37. Berthier, E., Young, E. W. K. & Beebe, D. Engineers are from PDMS-land, biologists are from polystyrenia. *Lab Chip* 12, 1224 (2012).

38. Wabuyele, M. B., Ford, S. M., Stryjewski, W., Barrow, J. & Soper, S. A. Single molecule detection of double-stranded DNA in poly(methylmethacrylate) and polycarbonate microfluidic devices. *Electrophoresis* 22, 3939–3948 (2001).

39. Chen, P. C., Park, D. S., You, B. H., Kim, N., Park, T., Soper, S. A., Nikitopoulos, D. E. & Murphy, M. C. Titer-plate formatted continuous flow thermal reactors: Design and performance of a nanoliter reactor. *Sens. Actuators, B* 149, 291–300 (2010).

40. Yang, W., Yu, M., Sun, X. & Woolley, A. T. Microdevices integrating affinity columns and capillary electrophoresis for multibiomarker analysis in human serum. *Lab Chip* 10, 2527–2533 (2010).

41. Peng, Z., Soper, S. A., Pingle, M. R., Barany, F. & Davis, L. M. Ligase detection reaction generation of reverse molecular beacons for near real-time analysis of bacterial pathogens using single-pair fluorescence resonance energy transfer and a cyclic olefin copolymer microfluidic chip. *Anal. Chem.* 82, 9727–9735 (2010).

42. Roy, E., Geissler, M., Galas, J.-C. & Veres, T. Prototyping of microfluidic systems using a commercial thermoplastic elastomer. *Microfluid. Nanofluidics* 11, 235–244 (2011); Brassard, D., Clime, L., Li, K., Geissler, M., Miville-Godin, C., Roy, E. & Veres, T. 3D thermoplastic elastomer microfluidic devices for biological probe immobilization. *Lab Chip* 11, 4099–4107 (2011).

43. Guillemette, M. D., Roy, E., Auger, F. A. & Veres, T. Rapid isothermal substrate microfabrication of a biocompatible thermoplastic elastomer for cellular contact guidance. *Acta Biomater.* 7, 2492–2498 (2011).

44. Roy, E., Galas, J.-C. & Veres, T. Thermoplastic elastomers for microfluidics: Towards a high-throughput fabrication method of multilayered microfluidic

devices. *Lab Chip 11*, 3193–3196 (2011); Didar, T. F., Li, K., Tabrizian, M. & Veres, T. High throughput multilayer microfluidic particle separation platform using embedded thermoplastic-based micropumping. *Lab Chip 13*, 2615–2622 (2013).

45. Olanrewaju, A., Beaugrand, M., Yafia, M. & Juncker, D. Capillary microfluidics in microchannels: From microfluidic networks to capillaric circuits. *Lab Chip 18*, 2323–2347 (2018).

46. Juncker, D., Schmid, H., Drechsler, U., Wolf, H., Wolf, M., Michel, B., Rooij, N. & Delamarche, E. Autonomous microfluidic capillary system. *Anal. Chem. 74*, 6139–6144 (2002); Zimmermann, M., Hunziker, P. & Delamarche, E. Capillary pumps for autonomous capillary systems. *Lab Chip 7*, 119–125 (2007).

47. Delamarche, E., Bernard, A., Schmid, H., Michel, B. & Biebuyck, H. Patterned delivery of immunoglobulins to surfaces using microfluidic networks. *Science 276*, 779–781 (1997).

48. *See* Juncker (2002) and Zimmermann (2007).

49. Olanrewaju, A., Beaugrand, M., Yafia, M. & Juncker, D. Capillary microfluidics in microchannels: From microfluidic networks to capillaric circuits. *Lab Chip 18*, 2323–2347 (2018); Safavieh, R. & Juncker, D. Capillarics: Pre-programmed, self-powered microfluidic circuits built from capillary elements. *Lab Chip 13*, 4180 (2013).

50. Safavieh, R. & Juncker, D. Capillarics: Pre-programmed, self-powered microfluidic circuits built from capillary elements. *Lab Chip 13*, 4180 (2013).

51. Olanrewaju, A. O., Ng, A., Decorwin-Martin, P., Robillard, A. & Juncker, D. Microfluidic capillaric circuit for rapid and facile bacteria detection. *Anal. Chem. 89*, 6846–6853 (2017).

52. Olanrewaju, A. O., Ng, A., Decorwin-Martin, P., Robillard, A. & Juncker, D. Microfluidic capillaric circuit for rapid and facile bacteria detection. *Anal. Chem. 89*, 6846–6853 (2017); Olanrewaju, A. O., Robillard, A., Dagher, M. & Juncker, D. Autonomous microfluidic capillaric circuits replicated from 3D-printed molds. *Lab Chip 16*, 3804–3814 (2016).

53. Tung, Y.-C., Hsiao, A. Y., Allen, S. G., Torisawa, Y., Ho, M. & Takayama, S. High-throughput 3D spheroid culture and drug testing using a 384 hanging drop array. *Analyst 136*, 473–478 (2011); Hsiao, A. Y., Tung, Y. C., Kuo, C. H., Mosadegh, B., Bedenis, R., Pienta, K. J. & Takayama, S. Micro-ring structures stabilize

microdroplets to enable long term spheroid culture in 384 hanging drop array plates. *Biomed. Microdevices* 14, 313–323 (2012); Hsiao, A. Y., Tung, Y.-C. C., Qu, X., Patel, L. R., Pienta, K. J. & Takayama, S. 384 hanging drop arrays give excellent Z-factors and allow versatile formation of co-culture spheroids. *Biotechnol. Bioeng.* 109, 1293–1304 (2012).

54. Walker, G. M. & Beebe, D. J. A passive pumping method for microfluidic devices. *Lab Chip* 2, 131–134 (2002).

55. Berthier, E. & Beebe, D. J. Flow rate analysis of a surface tension driven passive micropump. *Lab Chip* 7, 1475 (2007).

56. Meyvantsson, I., Warrick, J. W., Hayes, S., Skoien, A. & Beebe, D. J. Automated cell culture in high density tubeless microfluidic device arrays. *Lab Chip* 8, 717–724 (2008).

57. Burtis, C. A., Mailen, J. C., Johnson, W. F., Scott, C. D., Tiffany, T. O. & Anderson, N. G. Development of a miniature fast analyzer. *Clin. Chem.* 18, 753–761 (1972).

58. Schembri, C. T., Burd, T. L., Kopf-Sill, A. R., Shea, L. R. & Braynin, B. Centrifugation and capillarity integrated into a multiple analyte whole blood analyser. *J. Automat. Chem.* 17, 99–104 (1995).

59. Church, G. M. & Kieffer-Higgins, S. Multiplex DNA sequencing. *Science* 240, 185–188 (1988).

60. Capitalizing on the genome. *Nat. Genet.* 13, 1–5 (1996).

61. Madou, M. J. & Kellogg, G. J. LabCD: A centrifuge-based microfluidic platform for diagnostics. *Proc. SPIE* 3259, 80–93 (1998).

62. Duffy, D. C., Gillis, H. L., Lin, J., Sheppard, N. F. & Kellogg, G. J. Microfabricated centrifugal microfluidic systems: Characterization and multiple enzymatic assays. *Anal. Chem.* 71, 4669–4678 (1999); Siegrist, J., Amasia, M., Singh, N., Banerjee, D. & Madou, M. Numerical modeling and experimental validation of uniform microchamber filling in centrifugal microfluidics. *Lab Chip* 10, 876–886 (2010); Gorkin, R., Park, J., Siegrist, J., Amasia, M., Lee, B. S., Park, J.-M., Kim, J., Kim, H., Madou, M. & Cho, Y.-K. Centrifugal microfluidics for biomedical applications. *Lab Chip* 10, 1758 (2010).

63. Gorkin, R., Park, J., Siegrist, J., Amasia, M., Lee, B. S., Park, J.-M., Kim, J., Kim, H., Madou, M. & Cho, Y.-K. Centrifugal microfluidics for biomedical applications. *Lab Chip* 10, 1758 (2010).

64. Park, J. M., Cho, Y. K., Lee, B. S., Lee, J. G. & Ko, C. Multifunctional microvalves control by optical illumination on nanoheaters and its application in centrifugal microfluidic devices. *Lab Chip 7*, 557–564 (2007).

65. Andersson, P., Jesson, G., Kylberg, G., Ekstrand, G. & Thorsén, G. Parallel nanoliter microfluidic analysis system. *Anal. Chem. 79*, 4022–4030 (2007).

66. Noroozi, Z., Kido, H., Micic, M., Pan, H., Bartolome, C., Princevac, M., Zoval, J. & Madou, M. Reciprocating flow-based centrifugal microfluidics mixer. *Rev. Sci. Instrum. 80*, (2009).

67. Höfflin, J., Torres Delgado, S. M., Suárez Sandoval, F., Korvink, J. G. & Mager, D. Electrifying the disk: A modular rotating platform for wireless power and data transmission for Lab on a disk application. *Lab Chip 15*, 2584–2587 (2015).

68. Grumann, M., Steigert, J., Riegger, L., Moser, I., Enderle, B., Riebeseel, K., Urban, G., Zengerle, R. & Ducrée, J. Sensitivity enhancement for colorimetric glucose assays on whole blood by on-chip beam-guidance. *Biomed. Microdevices 8*, 209–214 (2006).

69. Lee, B. S., Lee, J. N., Park, J. M., Lee, J. G., Kim, S., Cho, Y. K. & Ko, C. A fully automated immunoassay from whole blood on a disc. *Lab Chip 9*, 1548–1555 (2009).

70. Smith, S., Mager, D., Perebikovsky, A., Shamloo, E., Kinahan, D., Mishra, R., Torres Delgado, S. M. et al. CD-based microfluidics for primary care in extreme point-of-care settings. *Micromachines 7*, 1–32 (2016).

71. Bhamla, M. S., Benson, B., Chai, C., Katsikis, G., Johri, A. & Prakash, M. Hand-powered ultralow-cost paper centrifuge. *Nat. Biomed. Eng. 1*, 1–7 (2017).

72. Liu, C. H., Chen, C. A., Chen, S. J., Tsai, T. T., Chu, C. C., Chang, C. C. & Chen, C. F. Blood plasma separation using a fidget-spinner. *Anal. Chem. 91*, 1247–1253 (2019).

73. Michael, I., Kim, D., Gulenko, O., Kumar, S., Kumar, S., Clara, J., Ki, D. Y. et al. A fidget spinner for the point-of-care diagnosis of urinary tract infection. *Nat. Biomed. Eng.* 1–10 (2020).

74. Batchelder, J. S. Dielectrophoretic manipulator. *Rev. Sci. Instrum. 54*, 300–302 (1983).

75. Colgate, E. & Matsumoto, H. An investigation of electrowetting-based microactuation. *J. Vac. Sci. Technol. A 8*, 3625 (1990).

76. Washizu, M. Electrostatic actuation of liquid droplets for microreactor applications. *IEEE Trans. Ind. App. 34*, 732–737 (1998).

77. Pollack, M. G., Fair, R. B. & Shenderov, A. D. Electrowetting-based actuation of liquid droplets for microfluidic applications. *Appl. Phys. Lett.* 77, 1725–1726 (2000).

78. Lee, J., Moon, H., Fowler, J., Schoellhammer, T. & Kim, C. J. Electrowetting and electrowetting-on-dielectric for microscale liquid handling. *Sens. Actuators, A* 95, 259–268 (2002).

79. Welch, E. R. F., Lin, Y. Y., Madison, A. & Fair, R. B. Picoliter DNA sequencing chemistry on an electrowetting-based digital microfluidic platform. *Biotechnol. J.* 6, 165–176 (2011).

80. Shamsi, M. H., Choi, K., Ng, A. H. C. & Wheeler, A. R. A digital microfluidic electrochemical immunoassay. *Lab Chip* 14, 547–54 (2014); Ng, A. H. C., Dean Chamberlain, M., Situ, H., Lee, V. & Wheeler, A. R. Digital microfluidic immunocytochemistry in single cells. *Nat. Commun.* 6, 7513 (2015).

81. Liu, Y. J., Yao, D. J., Lin, H. C., Chang, W. Y. & Chang, H. Y. DNA ligation of ultramicro volume using an EWOD microfluidic system with coplanar electrodes. *J. Micromechanics Microengineering* 18, (2008); Hua, Z., Rouse, J. L., Eckhardt, A. E., Srinivasan, V., Pamula, V. K., Schell, W. A., Benton, J. L., Mitchell, T. G. & Pollack, M. G. Multiplexed real-time polymerase chain reaction on a digital microfluidic platform. *Anal. Chem.* 82, 2310–2316 (2010); Chang, Y. H., Lee, G. Bin, Huang, F. C., Chen, Y. Y. & Lin, J. L. Integrated polymerase chain reaction chips utilizing digital microfluidics. *Biomed. Microdevices* 8, 215–225 (2006).

82. Barbulovic-Nad, I., Yang, H., Park, P. S. & Wheeler, A. R. Digital microfluidics for cell-based assays. *Lab Chip* 8, 519 (2008).

83. Ng, A. H. C., Li, B. B., Chamberlain, M. D. & Wheeler, A. R. Digital Microfluidic Cell Culture. *Annu. Rev. Biomed. Eng.* 17, 91–112 (2015).

84. Choi, K., Ng, A. H. C., Fobel, R. & Wheeler, A. R. Digital Microfluidics. *Annu. Rev. Anal. Chem.* 5, 413–440 (2012).

85. Arango, Y., Temiz, Y., Gökçe, O. & Delamarche, E. Electro-actuated valves and self-vented channels enable programmable flow control and monitoring in capillary-driven microfluidics. *Sci. Adv.* 6, eaay8305 (2020).

86. Byrne, J. D., Jajja, M. R. N., O'Neill, A. T., Bickford, L. R., Keeler, A. W., Hyder, N., Wagner, K. et al. Wireless optofluidic systems for programmable in vivo pharmacology and optogenetics. *Cell* 162, 662–674 (2015).

87. Caine, P. Northwestern engineering team pioneers new medical technologies. *WTTW*, September 9, 2019. https://news.wttw.com/2019/09/09/northwestern-engineering-team-pioneers-new-medical-technologies.

88. Rogers, J., Huang, Y., Schmidt, O. G. & Gracias, D. H. Origami MEMS and NEMS. *MRS Bull. 41*, 123–129 (2016).

89. Xu, S., Yan, Z., Jang, K.-I., Huang, W., Fu, H., Kim, J., Wei, Z. et al. Assembly of micro/nanomaterials into complex, three-dimensional architectures by compressive buckling. *Science 347*, 154–159 (2015).

90. Kim, D. H., Lu, N., Ma, R., Kim, Y.-S., Kim, R.-H., Wang, S., Wu, J. et al. Epidermal electronics. *Science 333*, 838–843 (2011).

91. Chung, H. U., Rwei, A. Y., Hourlier-Fargette, A., Xu, S., Lee, K., Dunne, E. C., Xie, Z. et al. Skin-interfaced biosensors for advanced wireless physiological monitoring in neonatal and pediatric intensive-care units. *Nat. Med. 26*, 418–429 (2020).

92. Viventi, J., Kim, D.-H., Moss, J. D., Kim, Y.-S., Blanco, J. A., Annetta, N., Hicks, A. et al. A conformal, bio-interfaced class of silicon electronics for mapping cardiac electrophysiology. *Sci. Transl. Med. 2*, 24ra22 (2010).

93. Kim, D.-H., Viventi, J., Amsden, J. J., Xiao, J., Vigeland, L., Kim, Y.-S., Blanco, J. A. et al. Dissolvable films of silk fibroin for ultrathin conformal bio-integrated electronics. *Nat. Mater. 9*, 511–517 (2010).

94. Kim, D.-H., Lu, N., Ghaffari, R., Kim, Y.-S., Lee, S. P., Xu, L., Wu, J. et al. Materials for multifunctional balloon catheters with capabilities in cardiac electrophysiological mapping and ablation therapy. *Nat. Mater. 10*, 316–323 (2011).

95. *See* Caine (2019).

96. Koh, A., Kang, D., Xue, Y., Lee, S., Pielak, R. M., Kim, J., Hwang, T. et al. A soft, wearable microfluidic device for the capture, storage, and colorimetric sensing of sweat. *Sci. Transl. Med. 8*, 366ra165 (2016).

97. Bandodkar, A. J., Gutruf, P., Choi, J., Lee, K., Skine, Y., Reeder, J. T., Jean, W. J. et al. Battery-free, skin-interfaced microfluidic/electronic systems for simultaneous electrochemical, colorimetric, and volumetric analysis of sweat. *Sci. Adv. 5*, eaav3294 (2019).

98. Kim, J., Banks, A., Xie, Z., Heo, S. Y., Gutruf, P., Lee, J. W., Xu, S. et al. Miniaturized flexible electronic systems with wireless power and near-field communication capabilities. *Adv. Funct. Mater. 25*, 4761–4767 (2015); Kim, S. B., Lee, K., Raj, M. S., Lee, B., Reeder, J. T., Koo, J., Hourlier-Fargette, A. et al. Soft, skin-interfaced microfluidic systems with wireless, battery-free electronics for digital, real-time tracking of sweat loss and electrolyte composition. *Small 14*, 1–9 (2018).

99. Choi, J., Kang, D., Han, S., Kim, S. B. & Rogers, J. A. Thin, Soft, skin-mounted microfluidic networks with capillary bursting valves for chrono-sampling of sweat. *Adv. Healthc. Mater.* *6*, 1601355 (2017).

100. Yu, Y., Nassar, J., Xu, C., Min, J., Yang, Y., Dai, A., Doshi, R. et al. Biofuel-powered soft electronic skin with multiplexed and wireless sensing for human-machine interfaces. *Sci. Robot.* 7946, 1–14 (2020).

101. Ray, T. R., Ivanovic, M., Curtis, P. M., Franklin, D., Guventurk, K., Jeang, W. J., Chafetz, J. et al. Soft, skin-interfaced sweat stickers for cystic fibrosis diagnostics and management. *Sci. Transl. Med.* *13*, eabd8109 (2021).

Chapter 5

1. Bonhoeffer, F. & Huf, J. In vitro experiments on axon guidance demonstrating an anterior-posterior gradient on the tectum. *EMBO J.* *1*, 427–431 (1982).

2. Walter, J., Kern-Veits, B., Huf, J., Stolze, B. & Bonhoeffer, F. Recognition of position-specific properties of tectal cell membranes by retinal axons in vitro. *Development* *101*, 685–696 (1987).

3. Vielmetter, J., Stolze, B., Bonhoeffer, F. & Stuermer, C. A. In vitro assay to test differential substrate affinities of growing axons and migratory cells. *Exp. Brain Res.* *81*, 283–287 (1990); Hornberger, M. R., Dütting, D., Ciossek, T., Yamada, T., Handwerker, C., Lang, S., Weth, F. et al. Modulation of EphA receptor function by coexpressed EphrinA ligands on retinal ganglion cell axons. *Neuron* *22*, 731–742 (1999).

4. Chaudhury, M. K. & Whitesides, G. M. Direct measurement of interfacial interactions between semispherical lenses and flat sheets of poly(dimethylsiloxane) and their chemical derivatives. *Langmuir* *7*, 1013 (1991).

5. Chaudhury, M. K. & Whitesides, G. M. Direct measurement of interfacial interactions between semispherical lenses and flat sheets of poly(dimethylsiloxane) and their chemical derivatives. *Langmuir* *7*, 1013 (1991); Ferguson, G. S., Chaudhury, M. K., Sigal, G. B. & Whitesides, G. M. Contact Adhesion of thin gold-films on elastomeric supports—Cold welding under ambient conditions. *Science* *253*, 776–778 (1991); Chaudhury, M. K. & Whitesides, G. M. Correlation between surface free-energy and surface constitution. *Science* *255*, 1230–1232 (1992); Chaudhury, M. K. & Whitesides, G. M. How to make water run uphill. *Science* *256*, 1539–1541 (1992); Ferguson, G. S., Chaudhury, M. K., Biebuyck, H. A. & Whitesides, G. M.

Monolayers on disordered substrates—Self-assembly of alkyltrichlorosilanes on surface-modified polyethylene and poly(dimethylsiloxane). *Macromolecules 26*, 5870–5875 (1993).

6. Abbott, N. L., Folkers, J. P. & Whitesides, G. M. Manipulation of the wettability of surfaces on the 0.1-micrometer to 1-micrometer scale through micromachining and molecular self-assembly. *Science 257*, 1380 (1992).

7. Kumar, A., Biebuyck, H. A., Abbott, N. L. & Whitesides, G. M. The use of self-assembled monolayers and a selective etch to generate patterned gold features. *J. Am. Chem. Soc. 114*, 9188 (1992).

8. Kumar, A. & Whitesides, G. M. Features of gold having micrometer to centimeter dimensions can be formed through a combination of stamping with an elastomeric stamp and an alkanethiol ink followed by chemical etching. *Appl. Phys. Lett. 63*, 2002 (1993).

9. Whitesides, G. M. The origins and the future of microfluidics. *Nature 442*, 368–373 (2006).

10. Jackman, R. J., Wilbur, J. L. & Whitesides, G. M. Fabrication of submicrometer features on curved substrates by microcontact printing. *Science 269*, 664–666 (1995) Jackman, R. J., Brittain, S. T., Adams, A., Prentiss, M. G. & Whitesides, G. M. Design and fabrication of topologically complex, three-dimensional microstructures. *Science 280*, 2089 (1998).

11. Wilbur, J. L., Jackman, R. J., Whitesides, G. M., Cheung, E. L., Lee, L. K. & Prentiss, M. G. Elastomeric optics. *Chem Mater 8*, 1380 (1996).

12. Folch, A., Wrighton, M. S. & Schmidt, M. A. Microfabrication of oxidation-sharpened silicon tips on silicon nitride cantilevers for atomic force microscopy. *J. Microelectromechanical Syst. 6*, 303 (1997).

13. Kim, E., Xia, Y. N. & Whitesides, G. M. Polymer microstructures formed by molding in capillaries. *Nature 376*, 581 (1995).

14. Folch, A. & Schmidt, M. A. Wafer-level in-registry microstamping. *J. Microelectromechanical Syst. 8*, 85–89 (1999).

15. Häussling, L., Michel, B., Ringsdorf, H. & Rohrer, H. Direct observation of streptavidin specifically adsorbed on biotin-functionalized self-assembled monolayers with the scanning tunneling microscope. *Angew. Chemie Int. Ed. English 30*, 569–572 (1991).

16. Anselmetti, D., Baratoff, A., Gunterodt, H.-J., Gerber, C., Michel, B. & Rohrer, H. Combined scanning tunneling and force microscopy. *J. Vac. Sci. Technol. B 12*, 1677 (1994).

17. Delamarche, E., Bernard, A., Schmid, H., Michel, B. & Biebuyck, H. Patterned delivery of immunoglobulins to surfaces using microfluidic networks. *Science 276*, 779–781 (1997).

18. Yalow, R. S. & Berson, S. A. Assay of plasma insulin in human subjects by immunological methods. *Nature 184*, 1648–1649 (1959).

19. Köhler, G. & Milstein, C. Continuous cultures of fused cells secreting antibody of predefined specificity. *Nature 256*, 495–497 (1975).

20. Effenhauser, C. S., Bruin, G. J. M., Paulus, A. & Ehrat, M. Integrated capillary electrophoresis on flexible silicone microdevices: Analysis of DNA restriction fragments and detection of single DNA molecules on microchips. *Anal. Chem. 69*, 3451–3457 (1997).

21. Folch, A. & Toner, M. Cellular micropatterns on biocompatible materials. *Biotechnol. Prog. 14*, 388–392 (1998).

22. Duffy, D. C., McDonald, J. C., Schueller, O. J. A. & Whitesides, G. M. Rapid prototyping of microfluidic systems in poly (dimethylsiloxane). *Anal. Chem. 70*, 4974–4984 (1998).

23. Lee, K. Y., LaBianca, N., Rishton, S. A., Zolgharnain, S., Gelorme, J. D., Shaw, J. & Chang, T. H. P. Micromachining applications of a high resolution ultrathick photoresist. *J. Vac. Sci. Technol. B 13*, 3012–3016 (1995).

24. Folch, A., Ayon, A., Hurtado, O., Schmidt, M. A. & Toner, M. Molding of deep polydimethylsiloxane microstructures for microfluidics and biological applications. *J. Biomech. Eng. 121*, 28 (1999).

25. Takayama, S., McDonald, J. C., Ostuni, E., Liang, M. N., Kenis, P. J. A., Ismagilov, R. F. & Whitesides, G. M. Patterning cells and their environments using multiple laminar fluid flows in capillary networks. *Proc. Natl. Acad. Sci. U. S. A. 96*, 5545 (1999); Takayama, S., Ostuni, E., LeDuc, P., Naruse, K., Ingber, D. E. & Whitesides, G. M. Subcellular positioning of small molecules. *Nature 411*, 1016 (2001).

26. Kenis, P. J. A., Ismagilov, R. & Whitesides, G. M. Microfabrication inside capillaries using multiphase laminar flow patterning. *Science 285*, 83 (1999).

27. Dertinger, S. K. W., Chiu, D. T., Jeon, N. L. & Whitesides, G. M. Generation of gradients having complex shapes using microfluidic networks. *Anal. Chem.* *73*, 1240–1246 (2001).

28. Dertinger, S. K. W., Jiang, X. Y., Li, Z. Y., Murthy, V. N. & Whitesides, G. M. Gradients of substrate-bound laminin orient axonal specification of neurons. *Proc. Natl. Acad. Sci. U. S. A.* *99*, 12542–12547 (2002).

29. Li Jeon, N., Baskaran, H., Dertinger, S. K. W., Whitesides, G. M., Van de Water, L. & Toner, M. Neutrophil chemotaxis in linear and complex gradients of interleukin-8 formed in a microfabricated device. *Nat. Biotechnol.* *20*, 826–830 (2002).

30. Keenan, T. M. & Folch, A. Biomolecular gradients in cell culture systems. *Lab Chip* *8*, 34–57 (2008).

31. Anderson, J. R., Chiu, D. T., Jackman, R. J., Cherniavskaya, O., McDonald, J. C., Wu, H. K., Whitesides, S. H. & Whitesides, G. M. Fabrication of topologically complex three-dimensional microfluidic systems in PDMS by rapid prototyping. *Anal. Chem.* *72*, 3158–3164 (2000).

32. Jo, B. H., Van Lerberghe, L. M., Motsegood, K. M. & Beebe, D. J. Three-dimensional micro-channel fabrication in polydimethylsiloxane (PDMS) elastomer. *J. Microelectromechanical Syst.* *9*, 76–81 (2000).

33. Chiu, D. T., Jeon, N. L., Huang, S., Kane, R. S., Wargo, C. J., Choi, I. S., Ingber, D. E. & Whitesides, G. M. Patterned deposition of cells and proteins onto surfaces by using three-dimensional microfluidic systems. *Proc. Natl. Acad. Sci. U. S. A.* *97*, 2408–13 (2000).

34. *See* Kenis (1999).

35. Quake, S. Solving the tyranny of pipetting. https://arxiv.org/abs/1802.05601 (2018).

36. Unger, M. A., Chou, H. P., Thorsen, T., Scherer, A. & Quake, S. R. Monolithic microfabricated valves and pumps by multilayer soft lithography. *Science* *288*, 113–116 (2000).

37. Popovic, Z. D., Sprague, R. A. & Neville Connell, G. A. Technique for monolithic fabrication of microlens arrays. *Appl. Opt.* *27*, 1281 (1988); Daly, D., Stevens, R. F., Hutley, M. C. & Davies, N. The manufacture of microlenses by melting photoresist. *Meas. Sci. Technol.* *1*, 759–766 (1990).

38. *See* Quake (2018).

39. Squires, T. M. & Quake, S. R. Microfluidics: Fluid physics at the nanoliter scale. *Rev. Mod. Phys.* 77, 977–1026 (2005).

40. Maerkl, S. J. & Quake, S. R. A systems approach to measuring the binding energy landscapes of transcription factors. *Science* 315, 233–237 (2007).

41. *See* Wilbur (1996).

42. Hsu, C. H. & Folch, A. Spatio-temporally-complex concentration profiles using a tunable chaotic micromixer. *Appl. Phys. Lett.* 89, 1–4 (2006).

43. Gonzalez-Suarez, A. M., Pena-Del Castillo, J. G., Hernández-Cruz, A. & Garcia-Cordero, J. L. Dynamic Generation of concentration- and temporal-dependent chemical signals in an integrated microfluidic device for single-cell analysis. *Anal. Chem.* 90, 8331–8336 (2018).

44. Lam, E. W., Cooksey, G. A., Finlayson, B. A. & Folch, A. Microfluidic circuits with tunable flow resistances. *Appl. Phys. Lett.* 89, 164105 (2006).

45. Chronis, N., Zimmer, M. & Bargmann, C. I. Microfluidics for in vivo imaging of neuronal and behavioral activity in Caenorhabditis elegans. *Nat. Methods* 4, 727–731 (2007).

46. Cooksey, G. A., Sip, C. G. & Folch, A. A multi-purpose microfluidic perfusion system with combinatorial choice of inputs, mixtures, gradient patterns, and flow rates. *Lab Chip* 9, 417–426 (2009).

47. de Hoyos-Vega, J. M., Gonzalez-Suarez, A. M. & Garcia-Cordero, J. L. A versatile microfluidic device for multiple ex vivo/in vitro tissue assays unrestrained from tissue topography. *Microsystems Nanoeng.* 6, (2020).

48. Yang, S.-Y., Lin, J.-L. & Lee, G.-B. A vortex-type micromixer utilizing pneumatically driven membranes. *J. Micromechanics Microengineering* 19, 035020 (2009).

49. Lai, H. & Folch, A. Design and dynamic characterization of "single-stroke" peristaltic PDMS micropumps. *Lab Chip* 11, 336–342 (2011).

50. Hosokawa, K. & Maeda, R. A pneumatically-actuated three-way microvalve fabricated with polydimethylsiloxane using the membrane transfer technique. *J. Micromechanics Microengineering* 10, 415 (2000).

51. Fu, A. Y., Chou, H. P., Spence, C., Arnold, F. H. & Quake, S. R. An integrated microfabricated cell sorter. *Anal. Chem.* 74, 2451–2457 (2002).

52. Liu, J., Hansen, C. & Quake, S. R. Solving the 'world-to-chip' interface problem with a microfluidic matrix. *Anal. Chem.* 75, 4718–4723 (2003).

53. Hong, J. W., Studer, V., Hang, G., Anderson, W. F. & Quake, S. R. A nanoliter-scale nucleic acid processor with parallel architecture. *Nat. Biotechnol.* 22, 435–439 (2004).

54. Marcus, J. S., Anderson, W. F. & Quake, S. R. Microfluidic single-cell mRNA isolation and analysis. *Anal. Chem.* 78, 3084–3089 (2006).

55. Kartalov, E. P. & Quake, S. R. Microfluidic device reads up to four consecutive base pairs in DNA sequencing-by-synthesis. *Nucleic Acids Res.* 32, 2873–2879 (2004).

56. Hansen, C. L., Skordalakes, E., Berger, J. M. & Quake, S. R. A robust and scalable microfluidic metering method that allows protein crystal growth by free interface diffusion. *Proc. Natl. Acad. Sci. U. S. A.* 99, 16531–16536 (2002); Hansen, C. L., Classen, S., Berger, J. M. & Quake, S. R. A microfluidic device for kinetic optimization of protein crystallization and in situ structure determination. *J. Am. Chem. Soc.* 128, 3142–3143 (2006).

57. Balagadde, F. K., You, L. C., Hansen, C. L., Arnold, F. H. & Quake, S. R. Long-term monitoring of bacteria undergoing programmed population control in a microchemostat. *Science* 309, 137–140 (2005).

58. Maerkl, S. J. & Quake, S. R. A systems approach to measuring the binding energy landscapes of transcription factors. *Science* 315, 233–237 (2007).

59. Ma, C., Fan, R., Ahmad, H., Shi, Q., Comin-Anduix, B., Chodon, T., Koya, R. C. et al. A clinical microchip for evaluation of single immune cells reveals high functional heterogeneity in phenotypically similar T cells. *Nat. Med.* 17, 738–743 (2011).

60. Beyer, S., Mohamed, T. & Walus, K. A microfluidics based 3D bioprinter with on-the-fly multimaterial switching capability. In *17th International Conference on Miniaturized Systems for Chemistry and Life Sciences*, pp. 176–178 (2013).

61. Delamarche, E., Schmid, H., Bietsch, A., Michel, B. & Biebuyck, H. Microfluidic networks for chemical patterning of substrates: Design and application to bioassays. *J. Am. Chem. Soc.* 120, 500 (1998).

62. Jo, B. H., Van Lerberghe, L. M., Motsegood, K. M. & Beebe, D. J. Three-dimensional micro-channel fabrication in polydimethylsiloxane (PDMS) elastomer. *J. Microelectromechanical Syst.* 9, 76–81 (2000); McDonald, J. C., Duffy, D. C.,

Anderson, J. R., Chiu, D. T., Wu, H. K., Schueller, O. J. A. & Whitesides, G. M. Fabrication of microfluidic systems in poly(dimethylsiloxane). *Electrophoresis 21*, 27–40 (2000).

63. Jo, B. H., Van Lerberghe, L. M., Motsegood, K. M. & Beebe, D. J. Three-dimensional micro-channel fabrication in polydimethylsiloxane (PDMS) elastomer. *J. Microelectromechanical Syst. 9*, 76–81 (2000).

64. Burger, B. V. & Munro, Z. Headspace gas analysis: Quantitative trapping and thermal desorption of volatiles using fused-silica open tubular capillary traps. *J. Chromatogr. 370*, 449 (1986); Baltussen, E., Sandra, P., David, F. & Cramers, C. Stir bar sorptive extraction (SBSE), a novel extraction technique for aqueous samples: Theory and principles. *J. Microcolumn Sep. 11*, 737–747 (1999); Tan, B. C. D., Marriott, P. J., Lee, K. & Morrison, P. D. Sorption of volatile organics from water using an open-tubular, wall-coated capillary column. *Analyst 125*, 469–475 (2000).

65. Toepke, M. W. & Beebe, D. J. PDMS absorption of small molecules and consequences in microfluidic applications. *Lab Chip 6*, 1484–1486 (2006).

66. Berthier, E., Young, E. W. K. & Beebe, D. Engineers are from PDMS-land, biologists are from polystyrenia. *Lab Chip 12*, 1224 (2012); Regehr, K. J., Domenech, M., Koepsel, J. T., Carver, K. C., Ellison-Zelski, S. J., Murphy, W. L., Schuler, L. A., Alarid, E. T. & Beebe, D. J. Biological implications of polydimethylsiloxane-based microfluidic cell culture. *Lab Chip 9*, 2132 (2009); Moore, T. A., Brodersen, P. & Young, E. W. K. Multiple myeloma cell drug responses differ in thermoplastic vs PDMS microfluidic devices. *Anal. Chem. 89*, 11391–11398 (2017); Halldorsson, S., Lucumi, E., Gómez-Sjöberg, R. & Fleming, R. M. T. Advantages and challenges of microfluidic cell culture in polydimethylsiloxane devices. *Biosens. Bioelectron. 63*, 218–231 (2015); Wang, J. D., Douville, N. J., Takayama, S. & ElSayed, M. Quantitative analysis of molecular absorption into PDMS microfluidic channels. *Ann. Biomed. Eng. 40*, 1862–1873 (2012); van Meer, B. J., de Vries, H., Firth, K. S. A., van Weerd, J., Tertoolen, L. G. J., Karperien, H. B. J., Jonkheijm, P., Denning, C., IJzerman, A. P. & Mummery, C. L. Small molecule absorption by PDMS in the context of drug response bioassays. *Biochem. Biophys. Res. Commun. 482*, 323–328 (2017); Shirure, V. S. & George, S. C. Design considerations to minimize the impact of drug absorption in polymer-based organ-on-a-chip platforms. *Lab Chip 17*, 681–690 (2017).

67. Gregory T. Roman, Hlaus, T., Bass, K. J., Seelhammer, T. G. & Culbertson, C. T. Sol–gel modified PDMS microfluidic devices with high electroosmotic mobilities and hydrophilic channel wall characteristics. *Anal. Chem. 77*, 1414–1422 (2005).

68. Li, M. & Kim, D. P. Silicate glass coated microchannels through a phase conversion process for glass-like electrokinetic performance. *Lab Chip 11*, 1126–1131 (2011).

69. Glick, C. C., Srimongkol, M. T., Schwartz, A. J., Zhuang, W. S., Lin, J. C., Warren, R. H., Tekell, D. R., Satamalee, P. A. & Lin, L. Rapid assembly of multilayer microfluidic structures via 3D-printed transfer molding and bonding. *Microsystems Nanoeng. 2*, 16063 (2016).

70. Hull, C. Stereolithography plastic prototypes from CAD data without tooling. *Mod. Cast. 78*, 38 (1988).

71. Bertsch, A., Bernhard, P., Vogt, C. & Renaud, P. Rapid prototyping of small size objects. *Rapid Prototyp. J. 6*, 259–266 (2000).

72. Bertsch, A., Heimgartner, S., Cousseau, P. & Renaud, P. Static micromixers based on large-scale industrial mixer geometry. *Lab Chip 1*, 56–60 (2001).

73. Guerin, L. J., Bossel, M., Demierre, M., Calmes, S. & Renaud, P. Simple and low cost fabrication of embedded micro-channels by using a new thick-film photoplastic. In *Proc., International Conference on Solid State Sensors and Actuators (Transducers '97)*, vol. 2, pp. 1419–1422 (1997); Renaud, P., van Lintel, H., Heuschkel, M. & Guerin, L. Photo-polymer microchannel technologies and applications. In *3rd International Symposium on Micro-Total Analysis Systems (MicroTAS'98, Banff, Canada)*, pp. 17–22. https:// doi:10.1007/978-94-011-5286-0 (1998); Heuschkel, M. O., Guerin, L., Buisson, B., Bertrand, D. & Renaud, P. Buried microchannels in photopolymer for delivering of solutions to neurons in a network. *Sens. Actuators, B B48*, 356–361 (1998).

74. EPFL. A "micro winery" that makes wine continuously. https://actu.epfl.ch/news/a-micro-winery-that-makes-wine-continuously/ (2016).

75. Au, A. K., Bhattacharjee, N., Horowitz, L. F., Chang, T. C. & Folch, A. 3D-printed microfluidic automation. *Lab Chip 15*, 1934–1941 (2015).

76. Rogers, C. I., Qaderi, K., Woolley, A. T. & Nordin, G. P. 3D printed microfluidic devices with integrated valves. *Biomicrofluidics 9*, 16501 (2015).

77. Rogers, C. I., Pagaduan, J. V, Nordin, G. P. & Woolley, A. T. Single-monomer formulation of polymerized polyethylene glycol diacrylate as a nonadsorptive material for microfluidics. *Anal. Chem. 83*, 6418–6425 (2011).

78. Rogers, C. I., Oxborrow, J. B., Anderson, R. R., Tsai, L. F., Nordin, G. P. & Woolley, A. T. Microfluidic valves made from polymerized polyethylene glycol diacrylate. *Sens. Actuators, B 191*, 438–444 (2014).

79. *See* Rogers (2011).

80. *See* Rogers (2015).

81. Xu, X. From cloud computing to cloud manufacturing. *Robot. Comput. Integr. Manuf. 28*, 75–86 (2012).

82. Lipson, H. & Kurman, M. *Fabricated: The New World of 3D Printing*. Wiley, 2013.

83. Ganguli, A., Mostafa, A., Berger, J., Aydin, M. Y., Sun, F., Ramirez, S. A. S. de, Valera, E., Cunningham, B. T., King, W. P. & Bashir, R. Rapid isothermal amplification and portable detection system for SARS-CoV-2. *Proc. Natl. Acad. Sci. U. S. A. 117* (2020).

84. Touchstone, L. A. Study: Portable, point-of-care COVID-19 test could bypass the lab. https://medicalxpress-com.cdn.ampproject.org/c/s/medicalxpress.com/news/2020-09-portable-point-of-care-covid-bypass-lab.amp (2020).

85. MakerBot. Thingiverse. http://www.thingiverse.com/.

86. Kitson, P. J., Rosnes, M. H., Sans, V., Dragone, V. & Cronin, L. Configurable 3D-Printed millifluidic and microfluidic 'lab on a chip' reactionware devices. *Lab Chip 12*, 3267 (2012); Kitson, P. J., Glatzel, S., Chen, W., Lin, C.-G., Song, Y.-F. & Cronin, L. 3D printing of versatile reactionware for chemical synthesis. *Nat. Protoc. 11*, 920–936 (2016); Symes, M. D., Kitson, P. J., Yan, J., Richmond, C. J., Cooper, G. J. T., Bowman, R. W., Vilbrandt, T. & Cronin, L. Integrated 3D-printed reactionware for chemical synthesis and analysis. *Nat. Chem.* 4, 349–354 (2012); Tsuda, S., Jaffery, H., Doran, D., Hezwani, M., Robbins, P. J., Yoshida, M. & Cronin, L. Customizable 3D printed 'plug and play' millifluidic devices for programmable fluidics. *PLoS One 10*, e0141640 (2015).

87. Capel, A. J., Edmondson, S., Christie, S. D. R., Goodridge, R. D., Bibb, R. J. & Thurstans, M. Design and additive manufacture for flow chemistry. *Lab Chip 13*, 4583 (2013).

88. Bishop, G. W., Satterwhite, J. E., Bhakta, S., Kadimisetty, K., Gillette, K. M., Chen, E. & Rusling, J. F. 3D-printed fluidic devices for nanoparticle preparation and flow-injection amperometry using integrated Prussian blue nanoparticle-modified electrodes. *Anal. Chem.* 87, 5437–5443 (2015); Carvajal, S., Fera, S. N., Jones, A. L., Baldo, T. A., Mosa, I. M., Rusling, J. F. & Krause, C. E. Disposable inkjet-printed electrochemical platform for detection of clinically relevant HER-2 breast cancer biomarker. *Biosens. Bioelectron. 104*, 158–162 (2018).

89. Li, F., Smejkal, P., Macdonald, N. P., Guijt, R. M. & Breadmore, M. C. One-step fabrication of a microfluidic device with an integrated membrane and embedded reagents by multimaterial 3D printing. *Anal. Chem.* *89*, 4701–4707 (2017).

90. Kotz, F., Mader, M., Dellen, N., Risch, P., Kick, A., Helmer, D. & Rapp, B. E. Fused deposition modeling of microfluidic chips in polymethylmethacrylate. *Micromachines 11*, 5–8 (2020).

Chapter 6

1. Wong, C. H., Siah, K. W., & Lo, A. W. Estimation of clinical trial success rates and related parameters. *Biostatistics 20*, 273–286 (2019).

2. Herper, M. The truly staggering costs of inventing new drugs. *Forbes* (2012).

3. *See* Wong (2019).

4. Tsukamoto, T. Animal disease models for drug screening: The elephant in the room? *Drug Discov. Today 21*, 529–530 (2016); Huan, J. Y., Miranda, C. L., Buhler, D. R. & Cheeke, P. R. Species differences in the hepatic microsomal enzyme metabolism of the pyrrolizidine alkaloids. *Toxicol. Lett. 99*, 127–137 (1998); Fashe, M. M., Juvonen, R. O., Petsalo, A., Räsänen, J. & Pasanen, M. Species-specific differences in the in vitro metabolism of lasiocarpine. *Chem. Res. Toxicol. 28*, 2034–2044 (2015); Sasahara, K., Shimokawa, Y., Hirao, Y., Koyama, N., Kitano, K., Shibata, M. & Umehara, K. Pharmacokinetics and metabolism of delamanid, a novel anti-tuberculosis drug, in animals and humans: Importance of albumin metabolism in vivo. *Drug Metab. Dispos. 43*, 1267–1276 (2015).

5. Shanks, N., Greek, R. & Greek, J. Are animal models predictive for humans? *Philos. Ethics, Humanit. Med. 4*, 2 (2009).

6. Dadgar, N., Gonzalez-Suarez, A. M., Fattahi, P., Hou, X., Weroha, J. S., Gaspar-maia, A., Stybayeva, G. & Revzin, A. A micro fluidic platform for cultivating ovarian cancer spheroids and testing their responses to chemotherapies. *Microsystems Nanoeng.* (2020) doi:10.1038/s41378-020-00201-6; Park, S. E., Georgescu, A. & Huh, D. Organoids-on-a-chip. *Science 965*, 960–965 (2019).

7. Huh, D., Fujioka, H., Tung, Y.-C., Futai, N., Paine, R., Grotberg, J. B. & Takayama, S. Acoustically detectable cellular-level lung injury induced by fluid mechanical stresses in microfluidic airway systems. *Proc. Natl. Acad. Sci. U. S. A. 104*, 18886–18891 (2007).

8. Huh, D., Matthews, B. D., Mammoto, A., Montoya-Zavala, M., Hsin, H. Y. & Ingber, D. E. Reconstituting organ-level lung functions on a chip. *Science 328*, 1662–1668 (2010).

9. Huh, D., Leslie, D. C., Matthews, B. D., Fraser, J. P., Jurek, S., Hamilton, G. A., Thorneloe, K. S., McAlexander, M. A. & Ingber, D. E. A human disease model of drug toxicity–induced pulmonary edema in a lung-on-a-chip microdevice. *Sci. Transl. Med. 4*, 159ra147 (2012).

10. Kim, H. J., Huh, D., Hamilton, G. & Ingber, D. E. Human gut-on-a-chip inhabited by microbial flora that experiences intestinal peristalsis-like motions and flow. *Lab Chip 12*, 2165–2174 (2012).

11. Cross, V. L., Zheng, Y., Won Choi, N., Verbridge, S. S., Sutermaster, B. A., Bonassar, L. J., Fischbach, C. & Stroock, A. D. Dense type I collagen matrices that support cellular remodeling and microfabrication for studies of tumor angiogenesis and vasculogenesis in vitro. *Biomaterials 31*, 8596–8607 (2010).

12. Hsu, Y. H., Moya, M. L., Hughes, C. C. W., George, S. C. & Lee, A. P. A microfluidic platform for generating large-scale nearly identical human microphysiological vascularized tissue arrays. *Lab Chip 13*, 2990–2998 (2013).

13. Homan, K. A., Gupta, N., Kroll, K. T., Kolesky, D. B., Skylar-Scott, M., Miyoshi, T., Mau, D. et al. Flow-enhanced vascularization and maturation of kidney organoids in vitro. *Nat. Methods 16*, 255–262 (2019).

14. Nikolaev, M., Mitrofanova, O., Broguiere, N., Geraldo, S., Dutta, D., Tabata, Y., Elci, B. et al. Homeostatic mini-intestines through scaffold-guided organoid morphogenesis. *Nature 585*, 574–578 (2020).

15. Viravaidya, K., Sin, A. & Shuler, M. L. Development of a microscale cell culture analog to probe naphthalene toxicity. *Biotechnol. Prog. 20*, 316–323 (2004).

16. Domansky, K., Inman, W., Serdy, J., Dash, A., Lim, M. H. M. & Griffith, L. G. Perfused multiwell plate for 3D liver tissue engineering. *Lab Chip 10*, 51–58 (2010).

17. van Midwoud, P. M., Merema, M. T., Verpoorte, E. & Groothuis, G. M. M. A microfluidic approach for in vitro assessment of interorgan interactions in drug metabolism using intestinal and liver slices. *Lab Chip 10*, 2778–2786 (2010).

18. Mathur, A., Loskill, P., Shao, K., Huebsch, N., Hong, S., Marcus, S. G., Marks, N. et al. Human iPSC-based cardiac microphysiological system for drug screening applications. *Scientific Reports 5*, 8883 (2015).

19. Marsano, A., Conficconi, C., Lemme, M., Occhetta, P., Gaudiello, E., Votta, E., Cerino, G., Redaelli, A. & Rasponi, M. Beating heart on a chip: A novel microfluidic platform to generate functional 3D cardiac microtissues. *Lab Chip 16*, 599–610 (2016).

20. Seo, J., Byun, W. Y., Alisafaei, F., Georgescu, A., Yi, Y.-S., Massaro-Giordano, M., Shenoy, V. B., Lee, V., Bunya, V. Y. & Huh, D. Multiscale reverse engineering of the human ocular surface. *Nat. Med.* 25, 1310–1318 (2019).

21. *See* Seo (2019).

22. Stoppini, L., Buchs, P.-A. & Muller, D. A simple method for organotypic cultures of nervous tissue. *J. Neurosci. Methods* 37, 173–182 (1991).

23. Chang, T. C., Mikheev, A. M., Huynh, W., Monnat, R. J., Rostomily, R. C. & Folch, A. Parallel microfluidic chemosensitivity testing on individual slice cultures. *Lab Chip* 14, 4540–4551 (2014); Folch, A., Monnat, R. J., Chang, C., Horowitz, L., Sip, C. G. & Rostomily, R. C. Microfluidic assay apparatus and methods of use. US Patent 9,518,977, filed October 19, 2013, and issued December 13, 2016.

24. Rodriguez, A., Horowitz, L. F., Castro, K., Yeung, R. S., Rostomily, R. C. & Folch, A. A microfluidic platform for the delivery of panels of drugs to live tumor slices. *Lab Chip* 20, 1658–1675 (2020).

25. Horowitz, L. F., Rodriguez, A. D., Dereli, Z., Lin, R., Mikheev, A. M., Monnat Raymond J., J., Folch, A. & Rostomily, R. C. Multiplexed drug testing of tumor slices using a microfluidic platform. *Nat. Precis. Oncol.* 4, 12 (2020).

26. Hsu, Y. H., Moya, M. L., Hughes, C. C. W., George, S. C. & Lee, A. P. A microfluidic platform for generating large-scale nearly identical human microphysiological vascularized tissue arrays. *Lab Chip* 13, 2990–2998 (2013); Shirure, V. S., Bi, Y., Curtis, M. B., Lezia, A., Goedegebuure, M. M., Goedegebuure, S. P., Aft, R., Fields, R. C. & George, S. C. Tumor-on-a-chip platform to investigate progression and drug sensitivity in cell lines and patient-derived organoids. *Lab Chip* 18, 3687–3702 (2018); Phan, D. T. T., Wang, X., Craver, B. M., Sobrino, A., Zhao, D., Chen, J. C., Lee, L. Y. N., George, S. C., Lee, A. P. & Hughes, C. C. W. A vascularized and perfused organ-on-a-chip platform for large-scale drug screening applications. *Lab Chip* 17, 511–520 (2017).

27. Aref, A. R., Campisi, M., Ivanova, E., Portell, A., Larios, D., Piel, B. P., Mathur, N. et al. 3D microfluidic *ex vivo* culture of organotypic tumor spheroids to model immune checkpoint blockade. *Lab Chip 18*, 3129–3143 (2018); Jenkins, R. W., Aref, A. R., Lizotte, P. H., Ivanova, E., Stinson, S., Zhou, C. W., Bowden, M. et al.

Ex vivo profiling of PD-1 blockade using organotypic tumor spheroids. *Cancer Discov.* *8*, 196–215 (2018).

28. Astolfi, M., Peant, B., Lateef, M. A., Rousset, N., Kendall-Dupont, J., Carmona, E., Monet, F. et al. Micro-dissected tumor tissues on chip: an ex vivo method for drug testing and personalized therapy. *Lab Chip 16*, 312–325 (2016); Horowitz, L. F., Rodriguez, A. D., Au-Yeung, A., Bishop, K. W., Barner, L. A., Mishra, G., Raman, A. et al. Microdissected "cuboids" for microfluidic drug testing of intact tissues. *Lab Chip 21*, 122 (2021).

Chapter 7

1. Leavitt, S. A. "A private little revolution": The home pregnancy test in American culture. *Bull. Hist. Med. 80*, 317–345 (2006).

2. Ghalioungui, P., Khalil, S. & Ammar, A. R. On an ancient Egyptian method of diagnosing pregnancy and determining foetal sex. *Med. Hist.* *7*, 241–246 (1963).

3. Ghalioungui, Khalil & Ammar, Ancient Egyptian method.

4. Olszynko-Gryn, J. The demand for pregnancy testing: The Aschheim-Zondek reaction, diagnostic versatility, and laboratory services in 1930s Britain. *Stud. Hist. Philos. Sci. Part C Stud. Hist. Philos. Biol. Biomed. Sci. 47*, 233–247 (2014).

5. Shapiro, H. A. & Zwarenstein, H. A rapid test for pregnancy on *Xenopus laevis*. *Nature 133*, 762 (1934).

6. Yong, E. How a frog became the first mainstream pregnancy test. *The Atlantic* (2017).

7. Yalow, R. S. & Berson, S. A. Assay of plasma insulin in human subjects by immunological methods. *Nature 184*, 1648–1649 (1959).

8. Bonhams. Crane, Margaret, inventor. The first home pregnancy test. https://www.bonhams.com/auctions/22407/lot/37/ (2015).

9. Romm, C. Before there were home pregnancy tests. *The Atlantic* (2015).

10. Washburn, E. W. The dynamics of capillary flow. *Phys. Rev. 17*, 273–283 (1921); Darcy, H. *Les Fontaines Publiques de la Ville de Dijon*. Hachette, 1856 ed., 2018.

11. Safavieh, R. & Juncker, D. Capillarics: Pre-programmed, self-powered microfluidic circuits built from capillary elements. *Lab Chip 13*, 4180 (2013).

12. Bloomstein, T. M. & Ehrlich, D. J. Laser-chemical three-dimensional writing for microelectromechanics and application to standard-cell microfluidics. *J. Vac. Sci. Technol. B 10*, 2671 (1992).

13. Yagoda, H. Applications of confined spot tests in analytical chemistry: Preliminary paper. *Ind. Eng. Chem.—Anal. Ed. 9*, 79–82 (1937).

14. FolgerLibrary. Syrup of violets and science. *YouTube*, https://www.youtube.com/watch?v=pdEbMBeoaa8 (2011).

15. Szabadváry, F. Indicators—A historical perspective. *J. Chem. Educ. 41*, 285–287 (1964).

16. Maumené, E. J. Sur un nouveau réactif pour distinguer la présence du sucre dans certains liquides. *J. Pharm. 17*, 368–370 (1850); Oliver, G. On bedside urinary tests: Detection of sugar in the urine by means of test papers. *Lancet 121*, 858–860 (1883).

17. Gutzeit, G. *Helv. Chim. Acta 12*, 829 (1929).

18. Feigl, F. *Qualitative Analyse mit Hilfe von Tüpfelreaktionen. Akademische Verlagsgesellschaft*. Akademische Verlagsgesell-schaft m.b.H, 1931.

19. Yagoda, H. Applications of confined spot tests in analytical chemistry: Preliminary paper. *Ind. Eng. Chem.—Anal. Ed. 9*, 79–82 (1937).

20. Muller, R. H. & Clegg, D. L. Automatic paper chromatography. *Anal. Chem. 21*, 1123 (1949).

21. Clegg, D. L. Paper chromatography. *Anal. Chem. 22*, 48–59 (1950).

22. Free, A. H., Adams, E. C., Kercher, M. L., Free, H. M. & Cook, M. H. Simple specific test for urine glucose. *Clin. Chem. 3*, 163–167 (1957).

23. Kohn, J. A rapid method of estimating blood-glucose ranges. *Lancet 273*, 119–121 (1957).

24. Rajendran, R. & Rayman, G. Point-of-care blood glucose testing for diabetes care in hospitalized patients: An evidence-based review. *J. Diabetes Sci. Technol. 8*, 1081–1090 (2014).

25. Jones, G. & Kraft, A. Corporate venturing: The origins of Unilever's pregnancy test. *Bus. Hist. 46*, 100–122 (2004).

26. *See* Jones (2004).

27. Moreno, M. D. L., Cebolla, Á., Munõz-Suano, A., Carrillo-Carrion, C., Comino, I., Pizarro, Á., León, F., Rodríguez-Herrera, A. & Sousa, C. Detection of gluten immunogenic peptides in the urine of patients with coeliac disease reveals transgressions in the gluten-free diet and incomplete mucosal healing. *Gut 66*, 250–257 (2017).

28. Carrio, A., Sampedro, C., Sanchez-Lopez, J. L., Pimienta, M. & Campoy, P. Automated low-cost smartphone-based lateral flow saliva test reader for drugs-of-abuse detection. *Sensors (Switzerland) 15*, 29569–29593 (2015).

29. Pacifici, R., Farré, M., Pichini, S., Ortuño, J., Roset, P. N., Zuccaro, P., Segura, J. & De La Torre, R. Sweat testing MDMA with the Drugwipe analytical device: A controlled study with two volunteers. *J. Anal. Toxicol. 25*, 144–146 (2001).

30. Magambo, K. A., Kalluvya, S. E., Kapoor, S. W., Seni, J., Chofle, A. A., Fitzgerald, D. W. & Downs, J. A. Utility of urine and serum lateral flow assays to determine the prevalence and predictors of cryptococcal antigenemia in HIV-positive outpatients beginning antiretroviral therapy in Mwanza, Tanzania. *J. Int. AIDS Soc. 17*, 1–6 (2014).

31. Schramm, E. C., Staten, N. R., Zhang, Z., Bruce, S. S., Kellner, C., Atkinson, J. P., Kyttaris, V. C. et al. A quantitative lateral flow assay to detect complement activation in blood. *Anal. Biochem. 477*, 78–85 (2015).

32. Schramm et al., Quantitative lateral flow assay; Ang, S. H., Rambeli, M., Thevarajah, T. M., Alias, Y. B. & Khor, S. M. Quantitative, single-step dual measurement of hemoglobin A1c and total hemoglobin in human whole blood using a gold sandwich immunochromatographic assay for personalized medicine. *Biosens. Bioelectron. 78*, 187–193 (2016).

33. Boisen, M. L., Oottamasathien, D., Jones, A. B., Millett, M. M., Nelson, D. S., Bornholdt, Z. A., Fusco, M. L. et al. Development of prototype filovirus recombinant antigen immunoassays. *J. Infect. Dis. 212*, S359–S367 (2015).

34. Nielsen, K., Yu, W. L., Kelly, L., Bermudez, R., Renteria, T., Dajer, A., Gutierrez, E., Williams, J., Algire, J. & De Eschaide, S. T. Development of a lateral flow assay for rapid detection of bovine antibody to *Anaplasma marginale*. *J. Immunoass. Immunochem. 29*, 10–18 (2008).

35. Rohrman, B. A., Leautaud, V., Molyneux, E. & Richards-Kortum, R. R. A lateral flow assay for quantitative detection of amplified HIV-1 RNA. *PLoS One 7*, (2012); Kamphee, H., Chaiprasert, A., Prammananan, T., Wiriyachaiporn, N.,

Kanchanatavee, A. & Dharakul, T. Rapid molecular detection of multidrug-resistant tuberculosis by PCR-nucleic acid lateral flow immunoassay. *PLoS One* *10*, 1–17 (2015).

36. Kim, Y. K., Lim, S. I., Cho, I. S., Cheong, K. M., Lee, E. J., Lee, S. O., Kim, J. B. et al. A novel diagnostic approach to detecting porcine epidemic diarrhea virus: The lateral immunochromatography assay. *J. Virol. Methods* *225*, 4–8 (2015).

37. Morales-Narváez, E., Naghdi, T., Zor, E. & Merkoçi, A. Photoluminescent lateral-flow immunoassay revealed by graphene oxide: Highly sensitive paper-based pathogen detection. *Anal. Chem.* *87*, 8573–8577 (2015).

38. Shyu, R. H., Shyu, H. F., Liu, H. W. & Tang, S. S. Colloidal gold-based immunochromatographic assay for detection of ricin. *Toxicon* *40*, 255–258 (2002).

39. Kuang, H., Xing, C., Hao, C., Liu, L., Wang, L. & Xu, C. Rapid and highly sensitive detection of lead ions in drinking water based on a strip immunosensor. *Sensors (Switzerland)* *13*, 4214–4224 (2013).

40. Rezaei, M., Bazaz, S. R., Zhand, S., Sayyadi, N., Jin, D., Stewart, M. P. & Warkiani, M. E. Point of care diagnostics in the age of COVID-19. *Diagnostics* *11*, 9 (2021).

41. CGTN America. George Whitesides: Simple Solutions. *YouTube*, https://www.youtube.com/watch?v=fd210-0aENY (2016).

42. Sia, S. K., Linder, V., Parviz, B. A., Siegel, A. & Whitesides, G. M. An integrated approach to a portable and low-cost immunoassay for resource-poor settings. *Angew. Chemie—Int. Ed.* *43*, 498–502 (2004).

43. Sia, S. K., Linder, V., Parviz, B. A., Siegel, A. & Whitesides, G. M. An integrated approach to a portable and low-cost immunoassay for resource-poor settings. *Angew. Chemie-International Ed.* *43*, 498–502 (2004); Linder, V., Sia, S. K. & Whitesides, G. M. Reagent-loaded cartridges for valveless and automated fluid delivery in microfluidic devices. *Anal. Chem.* *77*, 64–71 (2005).

44. Martinez, A. W., Phillips, S. T., Butte, M. J. & Whitesides, G. M. Patterned paper as a platform for inexpensive, low-volume, portable bioassays. *Angew. Chemie—Int. Ed.* *46*, 1318–1320 (2007).

45. Martinez, A. W., Phillips, S. T. & Whitesides, G. M. Three-dimensional microfluidic devices fabricated in layered paper and tape. *Proc. Natl. Acad. Sci. U. S. A.* *105*, 19606–19611 (2008).

46. Martinez, A. W., Phillips, S. T., Carrilho, E., Iii, S. W. T., Sindi, H., Whitesides, G. M., Thomas, S. W., Sindi, H. & Whitesides, G. M. Simple telemedicine for developing regions: Camera phones and paper-based microfluidic devices for real-time, off-site diagnosis. *Anal. Chem.* *80*, 3699–3707 (2008).

47. Cheng, C.-M., Martinez, A. W., Gong, J., Mace, C. R., Phillips, S. T., Carrilho, E., Mirica, K. A. & Whitesides, G. M. Paper-based ELISA. *Angew. Chemie-International Ed.* *49*, 4771–4774 (2010); Liu, X. Y., Cheng, C. M., Martinez, A. W., Mirica, K. A., Li, X. J., Phillips, S. T., Mascareñas, M. & Whitesides, G. M. A portable microfluidic paper-based device for ELISA. In *Proc., IEEE International Conference on Micro Electro Mechanical Systems*, pp. 75–78. https://doi:10.1109/MEMSYS.2011.5734365 (2011).

48. Martinez, A. W., Phillips, S. T., Nie, Z., Cheng, C.-M., Carrilho, E., Wiley, B. J. & Whitesides, G. M. Programmable diagnostic devices made from paper and tape. *Lab Chip* *10*, 2499–2504 (2010).

49. Martinez, R. V., Fish, C. R., Chen, X. & Whitesides, G. M. Elastomeric origami: Programmable paper-elastomer composites as pneumatic actuators. *Adv. Funct. Mater.* *22*, 1376–1384 (2012).

50. Lu, R., Shi, W., Jiang, L., Qin, J. & Lin, B. Rapid prototyping of paper-based microfluidics with wax for low-cost, portable bioassay. *Electrophoresis* *30*, 1497–1500 (2009); Carrilho, E., Martinez, A. W. & Whitesides, G. M. Understanding wax printing: A simple micropatterning process for paper-based microfluidics. *Anal. Chem.* *81*, 7091–7095 (2009).

51. Diagnostics For All. Our history. http://dfa.org/history (2020).

52. Dungchai, W., Chailapakul, O. & Henry, C. S. Electrochemical detection for paper-based microfluidics. *Anal. Chem.* *81*, 5821–5826 (2009).

53. Anderson, C. E., Buser, J. R., Fleming, A. M., Strauch, E. M., Ladd, P. D., Englund, J., Baker, D. & Yager, P. An integrated device for the rapid and sensitive detection of the influenza hemagglutinin. *Lab Chip* *19*, 885–896 (2019).

54. Yamada, K., Shibata, H., Suzuki, K. & Citterio, D. Toward practical application of paper-based microfluidics for medical diagnostics: state-of-the-art and challenges. *Lab Chip* *17*, 1206–1249 (2017); Li, X., Ballerini, D. R. & Shen, W. A perspective on paper-based microfluidics: Current status and future trends. *Biomicrofluidics* *6*, 11301 (2012); Yetisen, A. K., Akram, M. S. & Lowe, C. R. Paper-based microfluidic point-of-care diagnostic devices. *Lab Chip* *13*, 2210–2251 (2013); Hu, J., Wang, S., Wang, L.,

Li, F., Pingguan-Murphy, B., Lu, T. J. & Xu, F. Advances in paper-based point-of-care diagnostics. *Biosens. Bioelectron.* 54, 585–597 (2014).

55. Toley, B. J., Wang, J. A., Gupta, M., Buser, J. R., Lafleur, L. K., Lutz, B. R., Fu, E. & Yager, P. A versatile valving toolkit for automating fluidic operations in paper microfluidic devices. *Lab Chip* 15, 1432–1444 (2015).

56. Ng, K., Gao, B., Yong, K. W., Li, Y., Shi, M., Zhao, X., Li, Z. et al. Paper-based cell culture platform and its emerging biomedical applications. *Mater. Today* 20, 32–44 (2017).

Chapter 8

1. American Cancer Society. Cancer facts and figures 2020. https://www.cancer.org/research/cancer-facts-statistics/all-cancer-facts-figures/cancer-facts-figures-2020.html (2020).

2. Wang, Y., Zhou, Y. & Hu, Z. The functions of circulating tumor cells in early diagnosis and surveillance during cancer advancement. *J. Transl. Intern. Med.* 5, 135–138 (2017).

3. Cristofanilli, M., Budd, G. T., Ellis, M. J., Stopeck, A., Matera, J., Miller, M. C., Reuben, J. M. et al. Circulating tumor cells, disease progression, and survival in metastatic breast cancer. *N. Engl. J. Med.* 351, 781–791 (2004).

4. Cohen, S. J., Punt, C. J. A., Iannotti, N., Saidman, B. H., Sabbath, K. D., Gabrail, N. Y., Picus, J. et al. Relationship of circulating tumor cells to tumor response, progression-free survival, and overall survival in patients with metastatic colorectal cancer. *J. Clin. Oncol.* 26, 3213–3221 (2008).

5. De Bono, J. S., Scher, H. I., Montgomery, R. B., Parker, C., Miller, M. C., Tissing, H., Doyle, G. V., Terstappen, L. W. W. M., Pienta, K. J. & Raghavan, D. Circulating tumor cells predict survival benefit from treatment in metastatic castration-resistant prostate cancer. *Clin. Cancer Res.* 14, 6302–6309 (2008).

6. Aceto, N., Bardia, A., Miyamoto, D. T., Donaldson, M. C., Wittner, B. S. , Spencer, J. A., Yu, M. et al. Circulating tumor cell clusters are oligoclonal precursors of breast cancer metastasis. *Cell* 158, 1110–1122 (2014).

7. Allard, W. J., Matera, J., Miller, M. C., Repollet, M., Connelly, M. C., Rao, C., Tibbe, A. G. J., Uhr, J. W. & Terstappen, L. W. M. M. Tumor cells circulate in the peripheral blood of all major carcinomas but not in healthy subjects or patients with nonmalignant diseases. *Clin. Cancer Res.* 10, 6897–6904 (2004).

8. Nanou, A., Crespo, M., Flohr, P., De Bono, J. S. & Terstappen, L. W. M. M. Scanning electron microscopy of circulating tumor cells and tumor-derived extracellular vesicles. *Cancers (Basel)*. *10*, 1–17 (2018).

9. Yu, M., Bardia, A., Aceto, N., Bersani, F., Madden, M. W., Donaldson, M. C., Desai, R. et al. Ex vivo culture of circulating breast tumor cells for individualized testing of drug susceptibility. *Science 345*, 216–220 (2014).

10. Lynch, T. J., Bell, D. W., Sordella, R., Gurubhagavatula, S., Okimoto, R. A., Brannigan, B. W., Harris, P. L. et al. Activating mutations in the epidermal growth factor receptor underlying responsiveness of non-small-cell lung cancer to gefitinib. *N. Engl. J. Med. 350*, 2129–2139 (2004); Sordella, R., Bell, D. W., Haber, D. A. & Settleman, J. Gefitinib-sensitizing EGFR mutations in lung cancer activate anti-apoptotic pathways. *Science 305*, 1163–1167 (2004).

11. Nagrath, S., Sequist, L. V., Maheswaran, S., Bell, D. W., Irimia, D., Ulkus, L., Smith, M. R. et al. Isolation of rare circulating tumour cells in cancer patients by microchip technology. *Nature 450*, 1235–1239 (2007).

12. Maheswaran, S., Sequist, L. V, Nagrath, S., Ulkus, L., Brannigan, B., Collura, C. V., Inserra, E. et al. Detection of mutations in EGFR in circulating lung-cancer cells. *N. Engl. J. Med. 359*, 366–377 (2008).

13. Sarioglu, A. F., Aceto, N., Kojic, N., Donaldson, M. C., Zeinali, M., Hamza, B., Engstrom, A. et al. A microfluidic device for label-free, physical capture of circulating tumor cell clusters. *Nat. Methods 12*, 685–691 (2015).

14. Huang, R., Barber, T. A., Schmidt, M. A., Tompkins, R. G., Toner, M., Bianchi, D. W., Kapur, R. & Flejter, W. L. A microfluidics approach for the isolation of nucleated red blood cells (NRBCs) from the peripheral blood of pregnant women. *Prenat. Diagn. 28*, 892–899 (2008).

15. Huang, L. R., Cox, E. C., Austin, R. H. & Sturm, J. C. Continuous particle separation through deterministic lateral displacement. *Science 304*, 987–990 (2004).

16. Di Carlo, D. Inertial microfluidics. *Lab Chip 9*, 3038 (2009).

17. Levi, P. and Orsini, V. *Abruzzo forte e gentile: Impressioni d'occhio e di cuore (Scrittori abruzzesi e scritti sull'Abruzzo)* [Abbruzzo strong and gentle: Impressions from the eye and the heart (Writers from Abruzzo and Writings about Abruzzo)]. Libreria editrice A. Di Cioccio, 1976.

18. Segre, G. & Silberberg, A. Radial particle displacements in Poiseuille flow of suspensions. *Nature 189*, 209–210 (1961).

19. Ookawara, S., Higashi, R., Street, D. & Ogawa, K. Feasibility study on concentration of slurry and classification of contained particles by microchannel. *Chem. Eng. J.* *101*, 171–178 (2004); Sudarsan, A. P. & Ugaz, V. M. Multivortex micromixing. *Proc. Natl. Acad. Sci. U. S. A.* *103*, 7228–7233 (2006).

20. Dean, W. R. Fluid motion in a curved channel. *Proc. R. Soc. London Ser. A* *121*, 402–420 (1928).

21. Di Carlo, D., Edd, J. F., Humphry, K. J., Stone, H. A. & Toner, M. Particle segregation and dynamics in confined flows. *Phys. Rev. Lett.* *102*, 1–4 (2009); Amini, H., Lee, W. & Di Carlo, D. Inertial microfluidic physics. *Lab Chip* *14*, 2739–2761 (2014).

22. Di Carlo, D., Irimia, D., Tompkins, R. G. & Toner, M. Continuous inertial focusing, ordering, and separation of particles in microchannels. *Proc. Natl. Acad. Sci. U. S. A.* *104*, 18892–18897 (2007).

23. Di Carlo, D., Edd, J. F., Irimia, D., Tompkins, R. G. & Toner, M. Equilibrium separation and filtration of particles using differential inertial focusing. *Anal. Chem.* *80*, 2204–2211 (2008).

24. Ozkumur, E., Shah, A. M., Ciciliano, J. C., Emmink, B. L., Miyamoto, D. T., Brachtel, E., Yu, M. et al. Inertial focusing for tumor antigen-dependent and -independent sorting of rare circulating tumor cells. *Sci. Transl. Med.* *5*, 179ra47 (2013); Fachin, F., Spuhler, P., Martel-Foley, J. M., Edd, J. F., Barber, T. A., Walsh, J., Karabacak, M. et al. Monolithic chip for high-throughput blood cell depletion to sort rare circulating tumor cells. *Sci. Rep.* *7*, 10936 (2017).

25. Ozkumur, E., Shah, A. M., Ciciliano, J. C., Emmink, B. L., Miyamoto, D. T., Brachtel, E., Yu, M. et al. Inertial focusing for tumor antigen-dependent and -independent sorting of rare circulating tumor cells. *Sci. Transl. Med.* *5*, 179ra47 (2013).

26. Warkiani, M. E., Guan, G., Luan, K. B., Lee, W. C., Bhagat, A. A. S., Kant Chaudhuri, P., Tan, D. S.-W. et al. Slanted spiral microfluidics for the ultra-fast, label-free isolation of circulating tumor cells. *Lab Chip* *14*, 128–137 (2014).

27. Mach, A. J., Kim, J. H., Arshi, A., Hur, S. C. & Di Carlo, D. Automated cellular sample preparation using a centrifuge-on-a-chip. *Lab Chip* *11*, 2827 (2011); Hur, S. C., Mach, A. J. & Di Carlo, D. High-throughput size-based rare cell enrichment using microscale vortices. *Biomicrofluidics* *5*, 1–11 (2011); Sollier, E., Go, D. E., Che, J., Gossett, D. R., O'Byrne, S., Weaver, W. M., Kummer, N. et al. Size-selective collection of circulating tumor cells using Vortex technology. *Lab Chip* *14*, 63–77 (2014); Che, J., Yu, V., Garon, E. B., Goldman, J. W. & Di Carlo, D. Biophysical isolation

and identification of circulating tumor cells. *Lab Chip 17*, 1452–1461 (2017); Che, J., Yu, V., Dhar, M., Renier, C., Matsumoto, M., Heirich, K., Garon, E. B. et al. Classification of large circulating tumor cells isolated with ultra-high throughput microfluidic Vortex technology. *Oncotarget* 7, 12748–60 (2016).

28. Mishra, A., Dubash, T. D., Edd, J. F., Jewett, M. & Suhaas, G. Ultra-high throughput magnetic sorting of large blood volumes for epitope-agnostic isolation of circulating tumor cells. *Proc. Natl. Acad. Sci. U. S. A.* 1–35 (2020). https://doi:10.1073/pnas.2006388117.

Chapter 9

1. Lipson, H. & Kurman, M. *Fabricated—The New World of 3D Printing.* Wiley, 2013; Zastrow, M. The new 3D printing. *Nature 578*, 20–24 (2020).

2. Naderi, A., Bhattacharjee, N. & Folch, A. Digital manufacturing for microfluidics. *Annu. Rev. Biomed. Eng. 21*, 325–364 (2019).

3. Erkal, J. L., Selimovic, A., Gross, B. C., Lockwood, S. Y., Walton, E. L., McNamara, S., Martin, R. S. & Spence, D. M. 3D printed microfluidic devices with integrated versatile and reusable electrodes. *Lab Chip 14*, 2023–2032 (2014).

4. Sochol, R. D., Sweet, E., Glick, C. C., Venkatesh, S., Avetisyan, A., Ekman, K. F., Raulinaitis, A. et al. 3D printed microfluidic circuitry via multijet-based additive manufacturing. *Lab Chip 16*, 668–678 (2016).

5. Hertafeld, E., Zhang, C., Jin, Z., Jakub, A., Russell, K., Lakehal, Y., Andreyeva, K. et al. Multi-material three-dimensional food printing with simultaneous infrared cooking. *3D Print. Addit. Manuf. 6*, 13–19 (2019).

6. Kong, Y. L., Tamargo, I. A., Kim, H., Johnson, B. N., Gupta, M. K., Koh, T. W., Chin, H. A., Steingart, D. A., Rand, B. P. & McAlpine, M. C. 3D printed quantum dot light-emitting diodes. *Nano Lett. 14*, 7017–7023 (2014).

7. Sun, K., Wei, T. S., Ahn, B. Y., Seo, J. Y., Dillon, S. J. & Lewis, J. A. 3D printing of interdigitated Li-ion microbattery architectures. *Adv. Mater. 25*, 4539–4543 (2013).

8. Muth, J. T., Vogt, D. M., Truby, R. L., Mengüç, Y., Kolesky, D. B., Wood, R. J. & Lewis, J. A. Embedded 3D printing of strain sensors within highly stretchable elastomers. *Adv. Mater. 26*, 6307–6312 (2014).

9. Adams, J. J., Duoss, E. B., Malkowski, T. F., Motala, M. J., Ahn, B. Y., Nuzzo, R. G., Bernhard, J. T. & Lewis, J. A. Conformal printing of electrically small antennas on three-dimensional surfaces. *Adv. Mater. 23*, 1335–1340 (2011).

10. Joe Lopes, A., MacDonald, E. & Wicker, R. B. Integrating stereolithography and direct print technologies for 3D structural electronics fabrication. *Rapid Prototyp. J. 18*, 129–143 (2012).

11. Mannoor, M. S., Jiang, Z., James, T., Kong, Y. L., Malatesta, K. A., Soboyejo, W. O., Verma, N., Gracias, D. H. & McAlpine, M. C. 3D printed bionic ears. *Nano Lett. 13*, 2634–2639 (2013).

12. Kang, E., Jeong, G. S., Choi, Y. Y., Lee, K. H., Khademhosseini, A. & Lee, S. H. Digitally tunable physicochemical coding of material composition and topography in continuous microfibres. *Nat. Mater. 10*, 877–883 (2011); Yamada, M., Utoh, R., Ohashi, K., Tatsumi, K., Yamato, M., Okano, T. & Seki, M. Controlled formation of heterotypic hepatic micro-organoids in anisotropic hydrogel microfibers for long-term preservation of liver-specific functions. *Biomaterials 33*, 8304–8315 (2012); Kobayashi, A., Yamakoshi, K., Yajima, Y., Utoh, R., Yamada, M. & Seki, M. Preparation of stripe-patterned heterogeneous hydrogel sheets using microfluidic devices for high-density coculture of hepatocytes and fibroblasts. *J. Biosci. Bioeng. 116*, 761–767 (2013).

13. Beyer, S., Mohamed, T. & Walus, K. A microfluidics based 3D bioprinter with on-the-fly multimaterial switching capability. In *17th International Conference on Miniaturized Systems for Chemistry and Life Sciences*, pp. 176–178 (2013).

14. Johnson, B. N., Lancaster, K. Z., Hogue, I. B., Meng, F., Kong, Y. L., Enquist, L. W. & McAlpine, M. C. 3D printed nervous system on a chip. *Lab Chip 16*, 1393–1400 (2016).

15. Hardin, J. O., Ober, T. J., Valentine, A. D. & Lewis, J. A. Microfluidic printheads for multimaterial 3D printing of viscoelastic inks. *Adv. Mater. 27*, 3279–3284 (2015).

16. Lind, J. U., Busbee, T. A., Valentine, A. D., Pasqualini, F. S., Yuan, H., Yadid, M., Park, S. J. et al. Instrumented cardiac microphysiological devices via multimaterial three-dimensional printing. *Nat. Mater. 16*, 303–308 (2017).

17. Smay, J. E., Cesarano, J. & Lewis, J. A. Colloidal inks for directed assembly of 3-D periodic structures. *Langmuir 18*, 5429–5437 (2002); Smay, J. E., Gratson, G. M., Shepherd, R. F., Cesarano, J. & Lewis, J. A. Directed colloidal assembly of 3D periodic structures. *Adv. Mater. 14*, 1279–1283 (2002).

18. Ahn, B. Y., Duoss, E. B., Motala, M. J., Guo, X., Park, S.-I., Xiong, Y., Yoon, J., Nuzzo, R. G., Rogers, J. A. & Lewis, J. A. Omnidirectional printing of flexible, stretchable, and spanning silver microelectrodes. *Science 323*, 1590–1593 (2009).

19. Gregory M. Gratson, Xu, M. & Lewis, J. A. Direct writing of three-dimensional webs. *Nature 428*, 386 (2004).

20. Wu, W., Deconinck, A. & Lewis, J. A. Omnidirectional printing of 3D microvascular networks. *Adv. Mater. 23*, H178–H183 (2011); Homan, K. A., Kolesky, D. B., Skylar-Scott, M. A., Herrmann, J., Obuobi, H., Moisan, A. & Lewis, J. A. Bioprinting of 3D convoluted renal proximal tubules on perfusable chips. *Sci. Rep. 6*, 34845 (2016).

21. Wehner, M., Truby, R. L., Fitzgerald, D. J., Mosadegh, B., Whitesides, G. M., Lewis, J. A. & Wood, R. J. An integrated design and fabrication strategy for entirely soft, autonomous robots. *Nature 536*, 451–455 (2016).

22. Mosadegh, B., Kuo, C.-H., Tung, Y.-C., Torisawa, Y., Bersano-Begey, T., Tavana, H. & Takayama, S. Integrated elastomeric components for autonomous regulation of sequential and oscillatory flow switching in microfluidic devices. *Nat. Phys. 6*, 433–437 (2010).

23. Skylar-Scott, M. A., Uzel, S. G. M., Nam, L. L., Ahrens, J. H., Truby, R. L., Damaraju, S. & Lewis, J. A. Biomanufacturing of organ-specific tissues with high cellular density and embedded vascular channels. *Sci. Adv. 5*, eaaw2459 (2019).

24. Noor, N., Shapira, A., Edri, R., Gal, I., Wertheim, L. & Dvir, T. 3D printing of personalized thick and perfusable cardiac patches and hearts. *Adv. Sci. 6*, 1900344 (2019).

25. Jaffe-Hoffman, M. Israeli scientists "print" world's first 3D heart with human tissue. *The Jerusalem Post* (2019).

26. Grigoryan, B., Paulsen, S. J., Corbett, D. C., Sazer, D. W., Fortin, C. L., Zaita, A. J., Greenfield, P. T. et al. Multivascular networks and functional intravascular topologies within biocompatible hydrogels. *Science 364*, 458–464 (2019).

INDEX

Note: Figures and boxes are indicated by "f" and "b" respectively, following page numbers.